ENGINEERING OF POLYMERS AND CHEMICAL COMPLEXITY

Volume II: New Approaches, Limitations, and Control

ENGINEERING OF POLYMERS AND CHEMICAL COMPLEXITY

Volume II: New Approaches, Limitations, and Control

Edited by

Walter W. Focke, PhD and Hans-Joachim Radusch, PhD

Gennady E. Zaikov, DSc, and A. K. Haghi, PhD

Reviewers and Advisory Board Members

Apple Academic Press

TORONTO   NEW JERSEY

Apple Academic Press Inc. | Apple Academic Press Inc.
3333 Mistwell Crescent | 9 Spinnaker Way
Oakville, ON L6L 0A2 | Waretown, NJ 08758
Canada | USA

©2014 by Apple Academic Press, Inc.

First issued in paperback 2021

Exclusive worldwide distribution by CRC Press, a member of Taylor & Francis Group
No claim to original U.S. Government works

ISBN 13: 978-1-77463-096-9 (pbk)
ISBN 13: 978-1-926895-87-1 (hbk)

Library of Congress Control Number: 2014937489

Library and Archives Canada Cataloguing in Publication

Engineering of polymers and chemical complexity.

Includes bibliographical references and index.
Contents: Volume I. Current state of the art and perspectives/edited by LinShu Liu, PhD, and Antonio Ballada, PhD; Gennady E. Zaikov, DSc, and A. K. Haghi, PhD, Reviewers and Advisory Board Members -- Volume II. New approaches, limitations and control / edited by Walter W Focke, PhD and Prof. Hans-Joachim Radusch.

ISBN 978-1-926895-86-4 (v. 1: bound).--ISBN 978-1-926895-87-1 (v. 2: bound)
1. Polymers. 2. Polymerization. 3. Chemical engineering. 4. Nanocomposites (Materials).
I. Liu, LinShu, editor of compilation II. Ballada, Antonio, editor of compilation III. Focke, W. W. (Walter Wilhelm), editor of compilation IV. Radusch, Hans-Joachim, editor of compilation V. Title: Current state of the art and perspectives. VI. Title: New approaches, limitations and control.

TP156.P6E54 2014   668.9   C2014-901112-1

Apple Academic Press also publishes its books in a variety of electronic formats. Some content that appears in print may not be available in electronic format. For information about Apple Academic Press products, visit our website at **www.appleacademicpress.com** and the CRC Press website at **www.crcpress.com**

ABOUT THE EDITORS

Walter W. Focke

Professor Walter W. Focke obtained his bachelor and master's degrees in chemical engineering from the University of Pretoria, South Africa, and a PhD in polymer science and engineering from the Massachusetts Institute of Technology (MIT), Cambridge, Massachusetts, USA. He is a full professor in the Department of Chemical Engineering and Director of the Institute of Applied Materials at the University of Pretoria. He teaches materials science and engineering as well as phase equilibrium thermodynamics at the undergraduate level and polymer processing and polymer additive technology at the postgraduate level. Professor Focke is a registered professional engineer, a member of the American Chemical Society, the Polymer Processing Society, the International Pyrotechnics Society, and the South African Institute of Chemical Engineers. His research is focused on chemical product design with emphasis on carbon materials, polymer additive technology, pyrotechnics, and prophylactic malaria control. Various thermal analysis techniques are employed to characterize and control oxidative processes in pyrotechnics, biodiesel and materials such as polymers and graphite.

Professor Focke has published more than 90 papers in peer-reviewed journals. His current Scopus h-Index is 8, and his publications listed in Scopus have been cited more than 1200 times. He is a member of the editorial boards of the *Journal of Vinyl and Additive Technology,* the *International Journal of Adhesion & Adhesives,* and *International Polymer Processing.*

Hans-Joachim Radusch, PhD

Professor Hans-Joachim Radusch is a full professor of polymer engineering at the Center of Engineering Science at Martin Luther University, Halle-Saale, Germany. He is a world-renowned scientist in the field of chemistry, physics and mechanic of polymers, polymer blends, polymer composites, and nanocomposites. He has published several books and 500 original papers and reviews in above fields of science.

REVIEWERS AND ADVISORY BOARD MEMBERS

Gennady E. Zaikov, DSc

Gennady E. Zaikov, DSc, is Head of the Polymer Division at the N. M. Emanuel Institute of Biochemical Physics, Russian Academy of Sciences, Moscow, Russia, and Professor at Moscow State Academy of Fine Chemical Technology, Russia, as well as Professor at Kazan National Research Technological University, Kazan, Russia. He is also a prolific author, researcher, and lecturer. He has received several awards for his work, including the the Russian Federation Scholarship for Outstanding Scientists. He has been a member of many professional organizations and on the editorial boards of many international science journals.

A. K. Haghi, PhD

A. K. Haghi, PhD, holds a BSc in urban and environmental engineering from University of North Carolina (USA); a MSc in mechanical engineering from North Carolina A&T State University (USA); a DEA in applied mechanics, acoustics and materials from Université de Technologie de Compiègne (France); and a PhD in engineering sciences from Université de Franche-Comté (France). He is the author and editor of 65 books as well as 1000 published papers in various journals and conference proceedings. Dr. Haghi has received several grants, consulted for a number of major corporations, and is a frequent speaker to national and international audiences. Since 1983, he served as a professor at several universities. He is currently Editor-in-Chief of the *International Journal of Chemoinformatics and Chemical Engineering* and *Polymers Research Journal* and on the editorial boards of many international journals. He is a member of the Canadian Research and Development Center of Sciences and Cultures (CRDCSC), Montreal, Quebec, Canada.

CONTENTS

LIST OF CONTRIBUTORS

A. Yu. Bedanokov
D. I. Mendeleev Russian University for Chemical Technology, Moscow, Russia.
Correspondence Address: 360004, Nalchik, Chernyshevskogo str. 173.

K. S. Dibirova
Dagestan State Pedagogical University, Makhachkala 367003, Yaragskii str. 57, Russian Federation.

I. V. Dolbin
Kh. M. Berbekov Kabardino-Balkarian State University, KBR, Nal'chik 360004, Chernyshevskii str., 173, Russian Federation.

R. A. Dvorikova
Institution of Russian Academy of Sciences A. Nesmeyanov Institute of Organoelement Compounds, RAS, Vavilov St. 28, 119991 Moscow, Russian Federation.

Ali Gharieh
Polymer Technology Research Laboratory, Department of Applied Chemistry, Faculty of Chemistry,University of Tabriz, Iran.
Email: a.gharieh@tabrizu.ac.ir

Mahdi Hasanzadeh
University of Guilan, Rasht, Iran.
Department of Textile Engineering, University of Guilan, Rasht, Iran.
Department of Textile Engineering, Amirkabir University of Technology, Tehran, Iran.
Department of Chemical Engineering, Imam Hossein Comprehensive University, Tehran, Iran.
Email: m_hasanzadeh@aut.ac.ir
Tel.: +98-21-33516875; fax: +98-182-3228375

N. I. Hloba
Belorusian State Technological University, 13a Sverdlov Str., Minsk, 2006, Belarus.
E-mail: prok_nr@mail.by

Peter Jurkovic
VIPO, a.s., Partizanske, Slovakia.

A. R. Khokhlov
Institution of Russian Academy of Sciences A. Nesmeyanov Institute of Organoelement Compounds, RAS, Vavilov Str. 28, 119991 Moscow, Russian Federation.

Z. S. Klemenkova
Institution of Russian Academy of Sciences A. Nesmeyanov Institute of Organoelement Compounds, RAS, Vavilov St. 28, 119991 Moscow, Russian Federation.

Grigorij Kogan
Directorate Health, Directorate General for Research and Innovation, European Commission, B-1049, Brussels, Belgium.

Yu. V. Korshak
D. Mendeleev University for Chemical Technology of Russia, Miusskaya Pl. 9, 125047 Moscow, Russian Federation.

A. Kostopoulou
Institute of Electronic Structure & Laser (IESL) Foundation for Research & Technology–Hellas (FORTH)
P. O. Box 1385, Vassilika Vouton 711 10 Heraklion, Crete.

G. V. Kozlov
Kabardino-Balkarian State University, Nal'chik-360004, Chernyshevsky str., 173, Russian Federation.
Dagestan State Pedagogical University, Makhachkala 367003, Yaragskii str., 57, Russian Federation.
Kh.M. Berbekov Kabardino-Balkarian State University, KBR, Nal'chik 360004, Chernyshevskii str., 173, Russian Federation.

E. T. Krutko
Belorusian State Technological University, 13a Sverdlov Str., Minsk, 2006, Belarus.
E-mail: prok_nr@mail

A. Lappas
Institute of Electronic Structure & Laser (IESL) Foundation for Research & Technology–Hellas (FORTH)
P. O. Box 1385, Vassilika Vouton 711 10 Heraklion, Crete

Marian Lehocky
Tomas Bata University in Zlín, T.G.M. Sq. 5555, 760 01 Zlín, Czech Republic.

G. M. Magomedov
Dagestan State Pedagogical University, Makhachkala 367003, Yaragskii str., 57, Russian Federation.

Jan Matyasovsky
VIPO, a.s., Partizanske, Slovakia.

A. K. Mikitaev
Kabardino–Balkarian State University, Nalchik, Russia.
Correspondence Address: 360004, Nalchik, Chernyshevskogo street 173.
Kh. M. Berbekov Kabardino-Balkarian State University, KBR, Nal'chik 360004, Chernyshevskii st., 173, Russian Federation.

M. A. Mikitaev
L. Ya. Karpov Research Institute, Moscow, Russia.

Bentolhoda Hadavi Moghadam
University of Guilan, Rasht, Iran.
Department of Textile Engineering, University of Guilan, Rasht, Iran.
Department of Textile Engineering, Amirkabir University of Technology, Tehran, Iran.

Mohammad Hasanzadeh Moghadam Abatari
Department of Mathematics, Faculty of Mathematical Sciences, University of Guilan, Rasht, Iran.

L. N. Nikitin
Institution of Russian Academy of Sciences A. Nesmeyanov Institute of Organoelement Cjmpounds, RAS, Vavilov Str. 28, 119991 Moscow, Russian Federation.

Igor Novak
Polymer Institute, Slovak Academy of Sciences, 845 41 Bratislava 45, Slovakia.
N. R. Prokopchuk
Belorusian State Technological University, 13a Sverdlov Str., Minsk, 2006, Belarus.
E-mail: prok_nr@mail.by
A. L. Rusanov
Institution of Russian Academy of Sciences A. Nesmeyanov Institute of Organoelement Compounds, RAS, Vavilov Str. 28, 119991 Moscow, Russian Federation.

V. A. Shanditsev
Institution of Russian Academy of Sciences A. Nesmeyanov Institute of Organoelement Compounds, RAS, Vavilov Str. 28, 119991 Moscow, Russian Federation.

Ladislav Šoltés
Institute of Experimental Pharmacology and Toxicology, Slovak Academy of Sciences, SK-84104 Bratislava, Slovakia.
FAX (+421-2)-5477-5928
E-mail: ELM ladislav.soltes@savba.sk

T. M. Tamer
Polymer Materials Research Department, Advanced Technologies and New Materials Research Institute (ATNMRI), City of Scientific Research and Technological Applications (SRTA- City), New Borg El-Arab City 21934, Alexandria, Egypt.
E-mail: ttamer85@gmail.com

R. R. Usmanova
Ufa State technical university of aviation, Ufa, Bashkortostan, Russia.
E-mail: Usmanovarr@mail.ru

Alenka Vesel
Department of Surface Engineering, Plasma Laboratory, Jožef Stefan Institute,Jamova cesta 39, SI-1000, Ljubljana, Slovenia.

G. E. Zaikov
N. M. Emanuel Institute of Biochemical Physics of Russian Academy of Sciences, Moscow-119334, Kosygin st., 4, Russian Federation.
E-mail: chembio@sky.chph.ras.ru

V. M. Zelencovsky
Institute of fiziko-organic chemistry National Academy of Sciences of Belarus, 13 Surganova Str., Minsk, 220072, Belarus.
E-mail: ela_Krutko@mail.ru

LIST OF ABBREVIATIONS

AFD	Average fiber diameter
ALS	Amyotrophic lateral sclerosis
ANN	Artificial neural network
ANOVA	Analysis of variance
ARDS	Respiratory distress syndrome
CA	Contact angle
CCBB	Continuous Configurational Boltzmann Biased
CCD	Central composite design
CNS	Central nervous system
CTMP	Chemithermomechanical pulp
CVD	Chemical vapor deposition
DB	Degree of branching
DCSBD	Diffuse coplanar surface barrier discharge
DFT	Density functional theory
DFT	Density functional theory
FP	Ferrocene-containing polymers
HBPs	Hyperbranched polymers
HLB	Low hydrophilic/lipophilic balance
IBD	Inflammatory bowel disease
IMP	Integral membrane protein
IRMOFs	Isoreticular metal-organic frameworks
LDA	Local density approximation
LW/AB	Lifshitz-van der Waals/acid-base
MD	Molecular dynamics
MOF	Metal-organic frameworks
NEMD	Non-equilibrium molecular dynamics
NSAIDs	Non-steroidal anti-inflammatory drugs
PSM	Post-synthetic modification
ROS	Reactive oxygen species
RSM	Response surface methodology
RWFT	Random walks in fractal time
SAIA	Slovak Academic Information Agency
SBUs	Secondary building units
SEM	Scanning electron microscopy
SF	Synovial fluid
STM	Scanning tunneling microscope
TBMD	Tight bonding molecular dynamics
TEM	Transmission electron

PREFACE

In studies of polymers and chemical complexity, we have designed a broad spectrum of new polymeric materials with unique architectures significant for emerging technologies. New experimental techniques are presented and sophisticated instrumentation are introduced about phenomena occurring at polymer surfaces and interfaces and how polymers diffuse or fracture. These advances in the understanding of polymer systems and chemical complexity are highlighted in the second volume of this series.

In the first chapter, application of polymeric nanocomposites filled with nanoparticles are introduced in detail. Dendritic architectures, as highly branched and three-dimensional macromolecules that have unique chemical and physical properties, offer potential as the next great technological revolution. Chapter 2 gives a brief introduction to some of the structural properties and application of dendritic polymer in various fields. The focus of this chapter is a survey of multi-scale modeling and simulation techniques in hyperbranched polymer and dendrimers. Results of modeling and simulation calculations on dendritic architecture are also reviewed. Polymer nanocomposites are commonly defined as the combination of a polymer matrix and additives that have at least one dimension in the nanometer range. One of the most important fields which have gained an increasing interest in recent years is magnetic nanocomposites. In chapter 3, the basics of magnetic properties of materials are presented along with emulsion polymerization approach to magnetic latexes. In chapter 4, the fractal analysis of polymerization kinetics in nanofiller presence was performed. The influence of catalyst structural features on chemical reaction course was shown.

One of the current trends in the synthesis of polyimides is the creation of fusible and soluble in organic solvents materials. This allows extending the range of their practical use and refines the classical methods for thermoplastics. This problem is particularly relevant in cases when the traditional scheme of high temperature prepolymer conversion into the final polymer, which usually takes place in the final product, cannot be carried out due to thermal instability of the product elements. It is often achieved by the use of monomers (diamines and dianhydrides) with bulky side groups for the synthesis of polyimides. The synthesis of polymers from a mixture of several diamines and dianhydrides, and especially the synthesis of block copolyimides, represents wide opportunities of directed regulation of polyimides properties, including giving them solubility. One of the methods of the block copolymers synthesis is getting them on the basis of pre-synthesized oligomers with determination molecular weight and with different functional groups.

Therefore in chapter 5, the synthesis of poly (4,4'-dipheniloxide)pyromellit(amic acid) (PAA), fragmented by oligo(amic acid) (OAA), obtained by low-temperature polycondensation of 4,4'-diaminodiphenyl oxide and dianhydride 4,4'-diphenyl-1,5-diazobicyclo/3,3,0/octane-2,3,6,7-tetracarboxilic acid and its subsequent chemical imidization. In the opposition to the original poly-(4,4'-aminodiphenyl)-pyromellitimide (PI) synthesized block copolyimides (BSPI) have a solubility in polar aprotonic solvents. The parameter of BSPI conformation was calculated to explain its solubility.

Chapter 6 describes oxidation stress -source and effects- Hyaluronan origin, properties and functions, and finally thiol compounds as antioxidants preventing HA degradations under conditions of oxidation stress.

In chapter 7, new magnetic nanomaterials have been synthesized from ferrocene-containing polyphenylenes. Cyclotrimerization of 1,1'-diacetylferrocene by condensation reaction catalyzed by p-toluenesulfonic acid in the presence of triethyl orthoformiate both in solution and supercritical carbon dioxide in the temperature range of 70–200°C Highly branched ferrocene-containing polyphenylenes prepared by this procedure were used as precursors for preparing magnetic nanomaterials. This was achieved by thermal treatment of polyphenylenes in the range of 200–750°C. The emerging of crystal magnetite nanoparticles of magnetite with the average size of 6–22 nm distributed in polyconjugated carbonized matrix was observed due to crosslinking and thermal degradation of polyphenylene prepolymers. Saturation magnetization of such materials came up to 32 Gs·cm3/g in a filed of 2.5 kOe.

In chapter 8, synthesis and structural properties of MOFs are summarized and some of the key advances that have been made in the application of these nanoporous materials in textile fibers are highlighted.

In chapter 9, a study has been conducted to investigate the relationship between four electrospinning parameters (solution concentration, applied voltage, tip to collector distance, and volume flow rate) and electrospun PAN nanofiber mat properties such as average fiber diameter (AFD) and contact angle (CA).

In chapter 10, the influence of four electrospinning parameters, comprising solution concentration, applied voltage, tip to collector distance, and volume flow rate on the CA of the electrospun PAN nanofiber mat was carried out using response surface methodology (RSM) and artificial neural network (ANN). First, a central composite design (CCD) was used to evaluate main and combined effects of above parameters. Then, these independent parameters were fed as inputs to an ANN while the output of the network was the CA of electrospun fiber mat. Finally, the importance of each electrospinning parameters on the variation of CA of electrospun fiber mat was determined and comparison of predicted CA value using RSM and ANN are discussed.

In chapter 11, the influence of crystalline morphology on fractal space formation for nanocomposites polymer/organoclay is presented.

In chapter 12, the fractal model of coke residue formation for composites high density polyethylene/aluminum hydroxide is described.

In chapter 13, the structural model of nanocomposites Poly(Vinyl Chloride)/organoclay flame-resistance is studied.

Calculation of efficiency of sedimentation of dispersion particles in A Rotoklon on the basis of model of hydrodynamic interacting of phases is presented in chapter 14.

In this book, we first briefly review the structure, properties, and application of dendritic macromolecules in various fields. Next, molecular simulation techniques in hyperbranched polymer and dendrimers is reviewed. Lastly, we will survey the most characteristic and important recent examples in molecular simulation of dendritic architectures.

Chapter 15 is about hyaluronan; a harbinger of the status and functionality of the joint, and the chapter 16 studies the polyvinylchloride antibacterial pre-treated by barrier plasma.

This new volume provides a balance between materials science and mechanics aspects, basic and applied research, and high technology composite development.

— Walter W. Focke, PhD and Hans-Joachim Radusch, PhD

CHAPTER 1

POLYMERIC NANOCOMPOSITES: STRUCTURE, MANUFACTURE, AND PROPERTIES

A. K. MIKITAEV, A. YU. BEDANOKOV, and M. A. MIKITAEV

CONTENTS

1.1   INTRODUCTION

The polymeric nanocomposites are the polymers filled with nanoparticles which interact with the polymeric matrix on the molecular level in contrary to the macrointeraction in composite materials. Mentioned nanointeraction results in high adhesion hardness of the polymeric matrix to the nanoparticles [1,52].Usual nanoparticle is less than 100 nanometers in any dimension, 1 nanometer being the billionth part of a meter [1,2].

The analysis of the reported studies tells that the investigations in the field of the polymeric nanocomposite materials are very promising.

The first notion of the polymeric nanocomposites was given in patent in 1950 [3]. Blumstain pointed in 1961 [4] that polymeric clay – based nanocomposites had increased thermal stability. It was demonstrated using the data of the thermogravimetric analysis that the polymethylmetacrylate intercalated into the Na^+ - methylmetacrylate possessed the temperature of destruction 40–50°C higher than the initial sample.

This branch of the polymeric chemistry did not attract much attention until 1990 when the group of scientists from the Toyota Concern working on the polyamide – based nanocomposites [5-9] found two – times increase in the elasticity modulus using only 4.7 weight% of the inorganic compound and 100°C increase in the temperature of destruction, both discoveries widely extending the area of application of the polyamide. The polymeric nanocomposites based on the layered silicates began being intensively studied in state, academic and industrial laboratories all over the world only after that.

1.2   STRUCTURE OF THE LAYERED SILICATES

The study of the polymeric nanocomposites on the basis of the modified layered silicates (broadly distributed and well—known as various clays) is of much interest. The natural layered inorganic structures used in producing the polymeric nanocomposites are the montmorillonite [10-12], hectorite [13], vermiculite [14], kaolin, saponine [15], and others. The sizes of inorganic layers are about 220 and 1 nanometers in length and width respectively [16,17].

The perspective ones are the bentonite breeds of clays which include at least 70% of the minerals from the montmorillonite group.

Montmorillonite $(Na,K,Ca)(Al,Fe,Mg)[(Si,Al)_4O_{10}](OH)_2\Box nH_2O$, named after the province Montmorillion in France, is the high – dispersed layered aluminous silicate of white or gray color in which appears the excess negative charge due to the non-stoichiometric replacements of the cations of the crystal lattice, charge being balanced by the exchange cations from the interlayer space. The main feature of the montmorillonite is its ability to adsorb ions, generally cations, and to exchange them. It produces plastic masses with water and may enlarge itself 10 times. Montmorillonite enters the bentonite clays (the term "bentonite" is given after the place Benton in USA).

The inorganic layers of clays arrange the complexes with the gaps called layers or galleries. The isomorphic replacement within the layers (such as Mg^{2+} replacing Al^{3+} in octahedral structure or Al^{3+} replacing Si^{4+} in tetrahedral one) generates the negative charges which electrostatically are compensated by the cations of the alkali or alkali-earth metals located in the galleries (Figure 1) [18]. The bentonite is very hydrophilic because of this. The water penetrates the interlayer space of the montmorillonite, hydrates its surface and exchanges cations what results in the swelling of the bentonite. The further dilution of the bentonite in water results in the viscous suspension with bold tixotropic properties.

The more pronounced cation – exchange and adsorption properties are observed in the bentonites montmorillonite of which contains predominantly exchange cations of sodium.

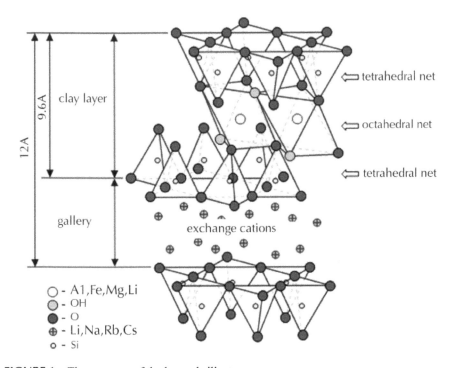

FIGURE 1   The structure of the layered silicate.

1.3   MODIFICATION OF THE LAYERED SILICATES

Layered silicates possess quite interesting properties – sharp drop of hardness at wetting, swelling at watering, dilution at dynamical influences, and shrinking at drying.

The hydrophility of aluminous silicates is the reason of their incompatibility with the organic polymeric matrix and is the first hurdle need to be overridden at producing the polymeric nanocomposites.

One way to solve this problem is to modify the clay by the organic substance. The modified clay (organoclay) has at least two advantages: (1) it can be well dispersed in polymeric matrix [19] and (2) it interacts with the polymeric chain [13].

The modification of the aluminous silicates can be done with the replacement of the inorganic cations inside the galleries by the organic ones. The replacing by the cationic surface – active agents like bulk ammonium and phosphonium ions increases the room between the layers, decreases the surface energy of clay and makes the surface of the clay hydrophobic. The clays modified such a way are more compatible with the polymers and form the layered polymeric nanocomposites [52]. One can use the non-ionic modifiers besides the organic ones which link themselves to the clay surface through the hydrogen bond. Organoclays produced with help of non-ionic modifiers in some cases become more chemically stable than the organoclays produced with help of cationic modifiers (Figure 2 (I)) [20].

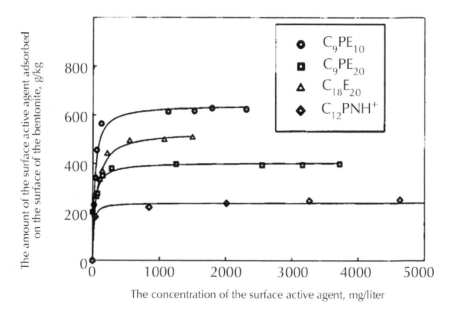

FIGURE 2 (I)   The adsorption of different modifiers on the clay surface.

The least degree of desorption is observed for non ionic interaction between the clay surface and organic modifier (Figure 2 (II)). The hydrogen bonds between the ethylenoxide grouping and the surface of the clay apparently make these organoclays more chemically stable than organoclays produced with non-ionic mechanism.

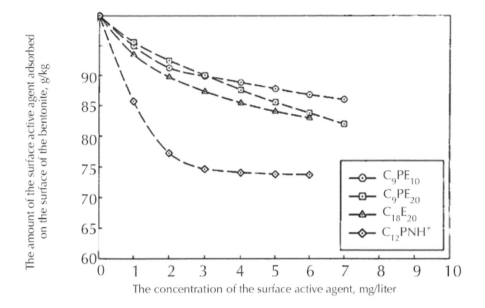

FIGURE 2 (II) The desorption of different modifiers from the clay surface:

$$C_9PE_{10} — C_9H_{19}C_6H_4(CH_2\ CH_2O)_{10}OH,$$
$$C_9PE_{20} — C_9H_{19}C_6H_4(CH_2\ CH_2O)_{20}OH,$$
$$C_{18}E_{20} — C_{18}H_{37}(CH_2\ CH_2O)_{20}OH,\ \text{and}$$
$$C_{12}PNH^+ — C_{12}H_{25}C_6H_4NH^+Cl^-.$$

1.4 STRUCTURE OF THE POLYMERIC NANOCOMPOSITES ON THE BASIS OF THE MONTMORILLONITE

The study of the distribution of the organoclay in the polymeric matrix is of great importance because the properties of composites obtained are in the direct relation from the degree of the distribution.

According Giannelis, the process of the formation of the nanocomposite goes in several intermediate stages (Figure 3) [21]. The formation of the tactoid happens on the first stage, the polymer surrounds the agglomerations of the organoclay. The polymer penetrates the interlayer space of the organoclay on the second stage. Here the gap between the layers may reach 2–3 nm [22]. The further separation of the layers, third stage, results in partial dissolution, and disorientation of the layers. Exfoliation is observed when polymer shifts the clay layers on more than 8–10 nm.

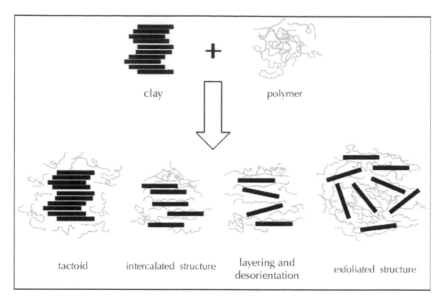

clay   polymer

tactoid   intercalated structure   layering and
desorientation   exfoliated structure

FIGURE 3   The schematic formation of the polymeric nanocomposite [24].

All mentioned structures may be present in real polymeric nanocomposites in dependence from the degree of distribution of the organoclay in the polymeric matrix. Exfoliated structure is the result of the extreme distribution of the organoclay. The excess of the organoclay or bad dispersing may born the agglomerates of the organoclays in the polymeric matrix what finds experimental confirmation in the X-ray analysis [11,12,21,23].

In the following subsections we describe a number of specific methods used at studying the structure of the polymeric nanocomposites.

1.4.1 DETERMINATION OF THE INTERLAYER SPACE

The X-ray determination of the interlayer distance in the initial and modified layered silicates as well as in final polymeric nanocomposite is one of the main methods of studying the structure of the nanocomposite on the basis of the layered silicate. The peak in the small angle diapason ($2\theta = 6–8°C$) is characteristic for pure clays and responds to the order of the structure of the silicate. This peak drifts to the smaller values of the angle 2θ in organomodified clays. If clay particles are uniformly distributed in the bulk of the polymeric matrix then this peak disappears, what witnesses on the disordering in the structure of the layered silicate. If the amount of the clay exceeds the certain limit of its distribution in the polymeric matrix, then the peak reappears again. This regularity was demonstrated on the instance of the polybutylenterephtalate (Figure 4) [11].

The knowledge of the angle 2θ helps to define the size of the pack of the aluminous silicate consisting of the clay layer and interlayer space. The size of such pack increases in a row from initial silicate to polymeric nanocomposite according to the increase in the interlayer space. The average size of that pack for montmorillonite is 1.2–1.5 nm but for organomodified one varies in the range of 1.8–3.5 nm.

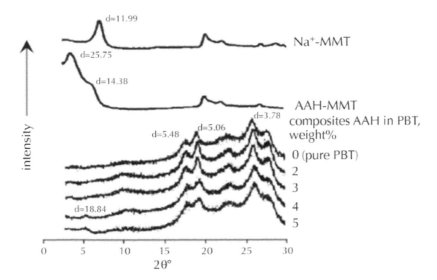

FIGURE 4 The data of the X-ray analysis for clay, organoclay, and nanocomposite PBT/organoclay.

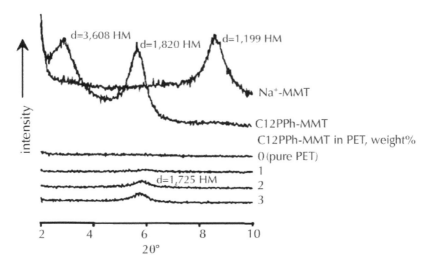

FIGURE 5 The data of the X-ray analysis for clay, organoclay, and nanocomposite PET/organoclay.

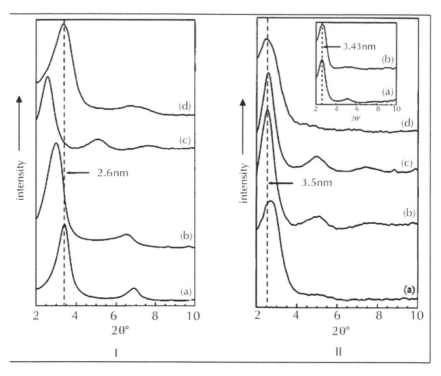

FIGURE 6   The data of the X-ray analysis for:

I. (a) Dimetyldioctadecylammonium (DMDODA)—hectorite,
    (b) 50% Polystyrene (PS)/50% DMDODA—hectorite,
    (c) 75% Polyethylmetacrylate (PEM)/25% DMDODA, and
    (d) 50% PS/50% DMDODA—hectorite after 24 hrs of etching in cyclohex-
       ane.
II. Mix of PS, PEM, and ofganoclay:
    (a) 23.8% PS/71.2% PEM/5% DMDODA—hectorite,
    (b) 21.2% PS/63.8% PEM/15% DMDODA—hectorite,
    (c) 18.2% PS/54.8% PEM/27% DMDODA—hectorite, and
    (d) 21.2% PS/63.8% PEM/15% DMDODA—hectorite after 24 hrs of etching
       in and cyclohexane.

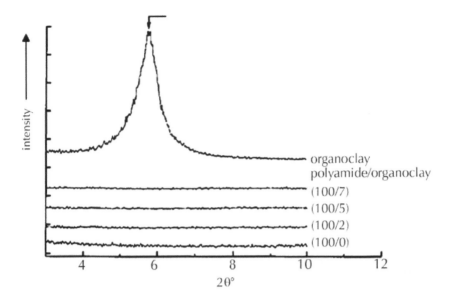

FIGURE 7 The data of the X-ray analysis for organoclay and nanocomposite polyamide acid/organoclay.

Summing up we conclude that comparing the data of the X-ray analysis for the organoclay and nanocomposite allows for the determination of the optimal clay amount need be added to the composite. The data from the scanning tunneling (STM) and transmission electron (TEM) microscopes [27,28] can be used as well.

1.4.2 THE DEGREE OF THE DISTRIBUTION OF THE CLAY PARTICLES IN THE POLYMERIC MATRIX

The two structures, namely the intercalated and exfoliated ones, could be distinguished with the respect to the degree of the distribution of the clay particles, Figure 8. One should note that clay layers are quite flexible though they are shown straight in the figure. The formation of the intercalated or exfoliated structures depends on many factors, for example the method of the production of the nanocomposite or the nature of the clay, and so on [29].

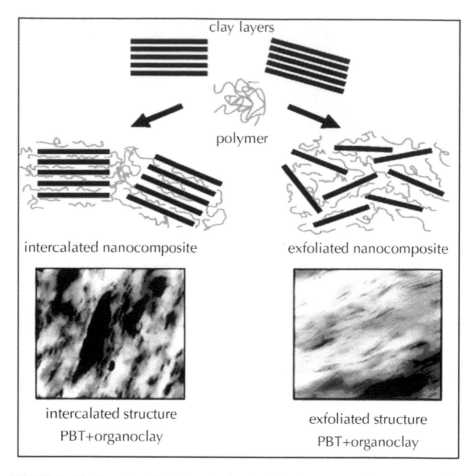

FIGURE 8   The formation of the intercalated and exfoliated structures of the nanocomposite.

The TEM images of the surface of the nanocomposites can help to find out the degree of the distribution of the nanosized clay particles, see plots (a) to (d) in the Figure 9.

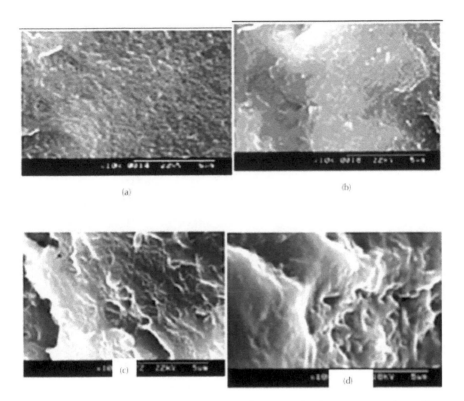

FIGURE 9   The images from scanning electron microscope for the nanocomposite surfaces:

(a) Pure PBT,
(b) 3 weight% of organoclay in PBT,
(c) 4 weight% of organoclay in PBT, and
(d) 5 weight% of organoclay in PBT.

The smooth surface tells about the uniform distribution of the organoclay particles. The surface of the nanocomposite becomes deformed with the increasing amount of the organoclay, see plots (a) to (d) in the Figure 10. Probably, this is due to the influence of the clay agglomerates [30,31].

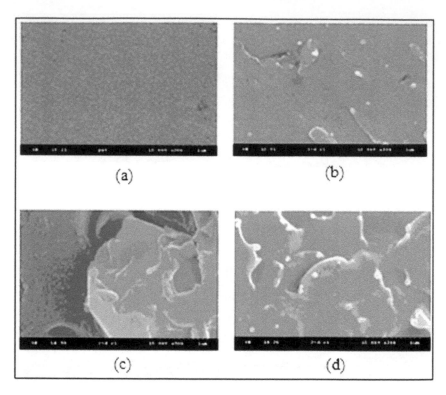

FIGURE 10   The images from scanning electron microscope for the nanocomposite surfaces:

(a) Pure PET,
(b) 3 weight% of organoclay in PET,
(c) 4 weight% of organoclay in PET, and
(d) 5 weight% of organoclay in PET.

Also one can use the STM images to judge on the degree of the distribution of the organoclay in the nanocomposite, Figures 11 and 12. If the content of the organoclay is 2–3weight% then the clay layers are separated by the polymeric layer of 4 to 10 nanometers width, Figure 11. If the content of the organoclay reaches 4–5weight% then the majority of the clay becomes well distributed however the agglomerates of 4–8 nanometers may appear.

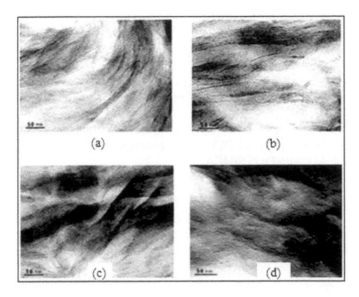

FIGURE 11   The images from tunneling electron microscope for the nanocomposite surfaces:

    (a) 2 weight% of organoclay in PBT,
    (b) 3 weight% of organoclay in PBT,
    (c) 4 weight% of organoclay in PBT, and
    (d) 5 weight% of organoclay in PBT.

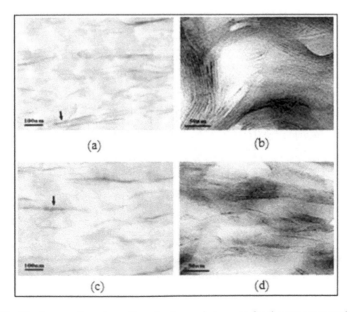

FIGURE 12   The images from tunneling electron microscope for the nanocomposite surfaces:

(a) 1 weight% of organoclay in PET,
(b) 2 weight% of organoclay in PET,
(c) 3 weight% of organoclay in PET, and
(d) 4 weight% of organoclay in PET.

So, the involvement of the X-ray analysis and the use of the microscopy data tell that the nanocomposite consists of the exfoliated clay at the low content (below 3 weight %) of the organoclay.

1.5 PRODUCTION OF THE POLYMERIC NANOCOMPOSITES ON THE BASIS OF THE ALUMINOUS SILICATES

The different groups of authors [32-35] offer following methods for obtaining nano-composites on the basis of the organoclays: (1) in the process of the synthesis of the polymer [33,36,37], (2) in the melt [38,39], (3) in the solution [40-46], and (4) in the sol-gel process [47-50].

The most popular ones are the methods of producing in melt and during the process of the synthesis of the polymer.

The producing of the polymeric nanocomposite in situ is the intercalation of the monomer into the clay layers. The monomer migrates through the organoclay galleries and the polymerization happens inside the layers [19,51], Figure 13.

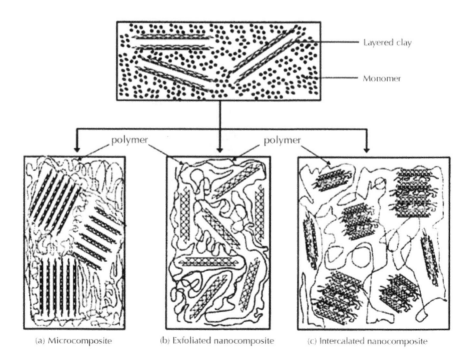

FIGURE 13   The production of the nanocomposite *in situ:*

(a) Microcomposite,
(b) Exfoliated nanocomposite, and
(c) Intercalated nanocomposite [51].

The polymerization may be initiated by the heat, irradiation or other source. Obviously, the best results on the degree of the distribution of the clay particles in the polymeric matrix must emerge if using given method. This is associated with the fact that the separation of the clay layers happens in the very process of the inclusion of the monomer in the interlayer space. In other words, the force responsible for the separation of the clay layers is the growth of the polymeric chain whereas the main factor for reaching the necessary degree of the clay distribution in solution or melt is just satisfactory mixing. The most favorable condition for synthesizing the nanocomposites is the vacuuming or the flow of the inert gas. Besides, one has to use the fast speeds of mixing for satisfactory dispersing of the organoclay in the polymeric matrix.

The method of obtaining the polymeric nanocomposites in melt (or the method of extrusion) is the mixing of the polymer melted with the organoclay. The polymeric chains lose the considerable amount of the conformational entropy during the intercalation. The probable motive force for this process is the important contribution of the enthalpy of the interaction between the polymer and organoclay at mixing. One should add that polymeric nanocomposite on the basis of the organoclays could be successively produced by the extrusion [22]. The advantage of the extrusion method is the absence of any solvents what excludes the unnecessary leaks. Moreover, the speed of the process is several times more and the technical side is simpler. The extrusion method is the best one in the industrial scales of production of the polymeric nanocomposites what acquires the lesser source expenses and easier technological scheme.

If one produces the polymer–silicate nanocomposite in solution then the organosilicate swells in the polar solvent such as toluene or N,N-dimethyl formamide. Then the added is the solution of the polymer which enters the interlayer space of the silicate. The removing of the solvent by means of evaporation in vacuum happens after that. The main advantage of the given method is that "polymer–layered silicate" might be produced from the polymer of low polarity or even non-polar one. However, this method is not widely used in industry because of much solvent consumption [52].

The sol-gel technologies find application at producing nanocomposites on the basis of the various ceramics and polymers. The initial compounds in these technologies are the alcoholates of specific elements and organic oligomers.

The alcoholates are firstly hydrolyzed and obtained hydroxides being polycondensated then. The ceramics from the inorganic 3D net is formed as a result. Also the method of synthesis exists in which the polymerization and the formation of the inorganic glass happen simultaneously. The application of the nanocomposites on the basis of ceramics and polymers as special hard defensive coverages and like optic fibers is possible [53].

1.6   PROPERTIES OF THE POLYMERIC NANOCOMPOSITES

Many investigations in physics, chemistry, and biology have shown that the jump from macroobjects to the particles of 1–10 nm results in the qualitative transformations in both separate phases and systems from them [54].

One can improve the thermal stability and mechanical properties of the polymers by inserting the organoclay particles into the polymeric matrix. It can be done by means of joining the complexes of properties of both the organic and inorganic substances that is combining the light weight, flexibility, and plasticity of former and durability, heat stability, and chemical resistance of latter.

Nanocomposites demonstrate essential change in properties if compared to the non-filled polymers. So, if one introduces modified layered silicates in the range of 2 to 10 weight% into the polymeric matrix then he observes the change in mechanical (tensile, compression, bending and overall strength), barrier (penetrability and stability to the solvent impact), optical and other properties. The increased heat and flame resistance even at low filler content is among the interesting properties too. The formation of the thermal isolation and negligible penetrability of the charred polymer to the flame provide for the advantages of using these materials.

The organoclay as a nanoaddition to the polymers may change the temperature of the destruction, refractoriness, rigidity, and rupture strength. The nanocomposites also possess the increased rigidity modulus, decreased coefficient of the heat expansion, low gas-penetrability, increased stability to the solvent impact and offer broad range of the barrier properties [54]. In Table 1 we gather the characteristics of the nylon-6 and its derivative containing 4.7 weight% of the organomodified montmorillonite.

TABLE 1   The properties of the nylon-6 and composite based on it [54]

	Rigidity modulus, GPa	Tensile strength, MPa	Temperature of the deformation, °C	Impact viscosity, kJ/m^2	Water consumption, weight %	Coefficient of the thermal expansion (x,y)
Nylon-6	1.11	68.6	65	6.21	0.87	13×10^{-5}
Nanocomposite	1.87	97.2	152	6.06	0.51	6.3×10^{-5}

It is important that the temperature of the deformation of the nanocomposite increases on 87°C.

The thermal properties of the polymeric nanocomposites with the varying organoclay content are collected in Table 2.

TABLE 2 The main properties of the polymeric nanocomposites

Property	Composition								
	Polybutyleneterephtalate + AAX-montmorillonite					Polyethyleneterephtalate + C_{12}PPh-montmorillonite			
	Organoclay content, %								
	0	2	3	4	5	0	1	2	3
Viscosity, dliter/g	0.84	1.16	0.77	0.88	0.86	1.02	1.26	0.98	1.23
T_g, °C	27	33	34	33	33	---	---	---	---
T_m, °C	222	230	230	229	231	245	247	245	246
T_d, °C	371	390	388	390	389	370	375	384	386
W_{tR}^{600c}, %	1	6	7	7	9	1	8	15	21
Strength limit, MPa	41	50	60	53	49	46	58	68	71
Rigidity modulus, GPa	1.37	1.66	1.76	1.80	1.86	2.21	2.88	3.31	4.10
Relative enlargement, %	5	7	6	7	7	3	3	3	3

The inclusion of the organoclay into the polybutyleneterephtalate leads to the increase in the glass transition temperature (T_g) from 27 to 33 degrees centigrade if the amount of the clay raises from zero to 2 weight%. That temperature does not change with the further increase of the organoclay content. The increase in T_g may be the result of two reasons [56-59]. The first is the dispersion of the small amount of the organoclay in the free volume of the polymer, and the second is the limiting of the mobility of the segments of the polymeric chain due to its interlocking between the layers of the organoclay.

The same as the T_g, the melting temperature T_m increases from the 222 to 230°C if the organoclay content raises from 0 to 2weight% and stays constant up to 5weight%, see Table 2. This increase might be the consequence of both complex multilayer structure of the nanocomposite and interaction between the organoclay and polymeric chain [60,61]. Similar regularities have been observed in other polymeric nanocomposites also.

The thermal stability of the nanocomposites polybutyleneterephtalate, briefly PBT, (or polyethyleleterephtalate)/organoclay determined by the thermogravimetric analysis is presented in Table 3 and in Figures 14 and 15 [11,12].

TABLE 3   The thermal properties of the fibers from PET with varying organoclay content

Organoclay content, weight %	$\eta_{inh}{}^a$	T_m (°C)	$\Delta H_m{}^b$ (J/g)	$T_D^{i\ c}$ (°C)	$Wt_R{}^{600d}$ (%)
0 (pure PET)	1.02	245	32	370	1
1	1.26	247	32	375	8
2	0.98	245	33	384	15
3	1.23	246	32	386	21

(a) Viscosities were measured at 30°C using 0.1 gram of polymer on 100 milliliters of solution in mix phenol/tetrachlorineethane (50/50),
(b) Change in enthalpy of melting,
(c) Initial temperature of decomposition, and
(d) Weight percentage of the coke remnant at 600°C.

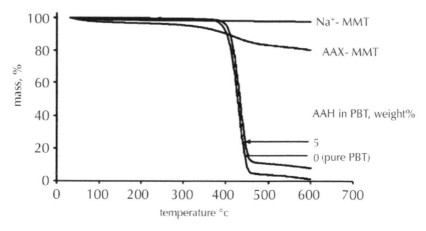

FIGURE 14 The thermogravimetrical curves for the montmorillonites, PBT, and nanocomposites PBT/organoclay.

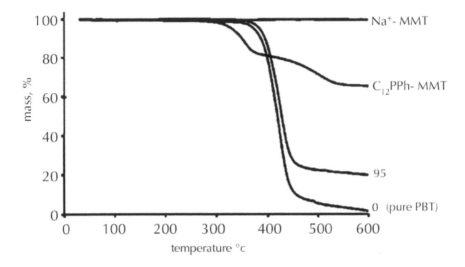

FIGURE 15 The thermogravimetrical curves for the montmorillonites, PET and nanocomposites PET/organoclay.

The temperature of the destruction, T_D, increases with the organoclay content up to 350°C in case of the composite PBT/organoclay. The thermogravimetrical curves for pure and composite PBTs have similar shapes below 350°C. The values of temperature T_D depends on the amount of organoclay above 350°C. The organoclay added becomes a barrier for volatile products being formed during the destruction [61,62]. Such example of the improvement of the thermal stability was studied in papers [63,64]. The mass of the remnant at 600°C increases with organoclay content.

Following obtained data authors draw the conclusion that the optimal results for thermal properties are being obtained if 2weight% of the organoclay is added [11,12,19].

The great number of studies on the polymeric composite organoclay–based materials show [11-13,19] that the inclusion of the inorganic component into the organic polymer improves the thermal stability of the latter, see Tables 3 and 4.

TABLE 4   The basic properties of the nanocomposite based on PBT with varying organoclay content

Organoclay content, weight %	I.V.[a]	T_g	T_m (°C)	$T_D^{i\ b}$ (°C)	Wt_R^{600c} (%)
0 (pure PBT)	0.84	27	222	371	1
2	1.16	33	230	390	6
3	0.77	34	230	388	7
4	0.88	33	229	390	7
5	0.86	33	231	389	9

(a) Viscosities were measured at 30°C using 0.1 gram of polymer on 100 milliliters of solution in mix phenol/tetrachlorineethane (50/50),
(b) Initial temperature of the weight loss, and
(c) Weight percentage of the coke remnant at 600°C.

The values of the melting temperature increase from 222 to 230°C if the amount of the organoclay added reaches 2 weight% and then stay constant. This effect can be explained by both thermal isolation of the clay and interaction between the polymeric chain and organoclay [43,64]. The increase in the glass transition temperature also occurs what can be a consequence of several reasons [55,56,58]. One of the main among them is the limited motion of the segments of the polymeric chain in the galleries within the organoclay.

If the organoclay content in the polymeric matrix of the PBT reaches 2 weight% then both the temperature of the destruction increases and the amount of the coke remnant increase at 660°C and then both stay practically unchanged with the further increase of the organoclay content up to 5 weight%. The loss of the weight due to the destruction of the polymer in pure PBT and its composites looks familiar in all cases below 350°C. The amount of the organoclay added becomes important above that temperature because the very clays possess good thermal stability and make thermal protection by their layers and form a barrier preventing the volatile products of the decomposition to fly off [43,60]. Such instance of the improvement in thermal properties was observed in many polymeric composites [64-68]. The weight of the coke remnant increases with the rising organoclay content up to 2 weight% and stays constant after that. The increase of the remnant may be linked with the high thermal stability of the organoclay itself. Also, it is worth noticing that the polymeric chain closed in interlayer space of the organoclay has fewer degrees of oscillatory motion at heating due to the limited interlayer space and the formation of the abundant intermolecular bonds between the polymeric chain and the clay surface. And the best result is obtained at 2–3 weight% content of the organoclay added to the polymer.

If one considers the influence of the organoclay added to the polyetheleneterephtalate, briefly PET, [69] then the temperature of the destruction increases on 16°C at optimal amount of organoclay of 3 weight%. The coke remnant at 600°C again increases with the rising organoclay content, see Table 3.

Regarding the change in the temperature of the destruction in the cases of PET and PBT versus the organoclay content one can note that both trends look similar. However the coke remnant considerably increases would the tripheyldodecylphosphonium cation be present within the clay. The melting temperature does not increase in case of the organoclay added into the PET in contrary to the case of PBT. Apparently this may be explained by the more crystallinity of the PBT and the growth of the degree of crystallinity with the organoclay content.

It becomes obvious after analyzing the above results that the introduction of the organoclay into the polymer increases the thermal stability of the latter according the (1) thermal isolating effect from the clay layers and (2) barrier effect in relation to the volatile products of destruction.

The studying of the mechanical properties of the nanocomposites, see Table 2, have shown that the limit of the tensile strength increases with the organoclay added up to 3 weight% for the majority of the composites. Further addition of the organoclay, up to 5 weight%, results in the decreasing limit of the tensile strength. We explain this by the fact that agglomerates appear in the nanocomposite when the organoclay content exceeds the 3 weight% value [61,70,71]. The proof for the formation of the agglomerates have been obtained from the X-ray study and using the data from electron microscopes.

Nevertheless, the rigidity modulus increases with the amount of the organoclay added into the polymeric matrix, the resistance of the clay itself being the explanation for that. The oriented polymeric chains in the clay layers also participate in the increase of the rigidity modulus [72]. The percentage of enlargement at breaking became 6–7 weight% for all mixes.

Using data of the Table 2 we explain the improvement in the mechanical properties of the nanocomposites with added organoclay up to 3 weight% by the good degree of distribution of the organoclay within the polymeric matrix. The degree of the improvement also depends on the interaction between the polymeric chain and clay layers.

The study of the influence of the degree of the extract of fibers on the mechanical properties has shown that the limit of strength and the rigidity modulus both increase in PBT whereas they decrease in nanocomposites, Table 5. This can be explained by the breaking of the bonds between the organoclay and PBT at greater degree of extract. Such phenomena have been observed in numerous polymeric composites [73-75].

TABLE 5   The ability to stretch of the nanocomposites PBT = organoclay at varying degrees of extract

Organoclay content,	Limit of strength, MPa			Rigidity modulus, GPa		
weight %	DR=1	DR=3	DR=6	DR=1	DR=3	DR=6
0 (pure PBT)	41	50	52	1.37	1.49	1.52
3	60	35	29	1.76	1.46	1.39

The first notions on the lowered flammability of the polymeric nanocomposites on the organoclay basis appeared in 1976 in the patent on the composite based on the nylon-6 [5]. The serious papers in the field were absent till the 1995 [76].

The use of the calorimeter is very effective for studying the refractoriness of the polymers. It can help at measuring the heat release, the carbon monoxide depletion and others. The speed of the heat release is one of the most important parameters defining the refractoriness [77]. The data on the flame resistance in various polymer / organoclay systems such as layered nanocomposite nylon-6/organoclay, intercalated nanocomposites polystyrene (or polypropylene)/organoclay were given in chapter [78] in where the lowered flammability was reported, see Table 6. And the lowered flammability have been observed in systems with low organoclay content, namely in range from 2 to 5weight%.

TABLE 6   Calorimetric data

Sample	remnant (%)±0.5	Peak of the HRR (Δ%) (kW/m²)	Middle of the HRR (Δ%) (kW/m²)	Average value H_c (MJ/ kg)	Average value SEA (m²/ kg)	Average CO left (kg/kg)
Nylon-6	1	1010	603	27	197	0.01
Nylon-6/organoclay, 2%, delaminated	3	686 (32%)	390 (35%)	27	271	0.01
Nylon-6/organoclay, 5%, delaminated	6	378 (63%)	304 (50%)	27	296	0.02
Polystyrene	0	1120	703	29	1460	0.09
PS/organoclay, 3%, bad mixing	3	1080	715	29	1840	0.09

TABLE 6   *(Continued)*

PS/organoclay, 3%, intercalated/delaminated	4	567 (48%)	444 (38%)	27	1730	0.08
PS w/DBDPO/ Sb$_2$O$_3$, 30%	3	491 (56%)	318 (54%)	11	2580	0.14
Polypropylene	0	1525	536	39	704	0.02
PP / organoclay, 2%, intercalated	5	450 (70%)	322 (40%)	44	1028	0.02

H$_c$——Heat of combustion;
SEA——Specific extinguishing area;
DBDPO——Dekabrominediphenyloxide;
HRR——Speed of the heat release

The curve of the heat release for the polypropylene and the nanocomposite on its basis (organoclay content varying from 2 to 4 weight%) is given in the Figure 16 from which one can see that the speed of the heat release for the nanocomposite enriched with the 4 weight% organoclay (the interlayer distance 3.5 nanometers) is 75% less than for pure polypropylene.

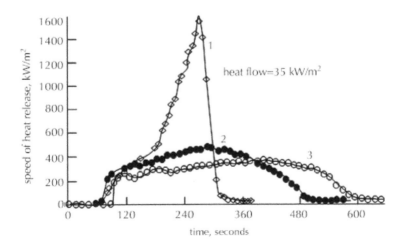

FIGURE 16   The speed of the heat release for:

1—Pure polypropylene,
2—Nanocomposite with 2 weight% of organoclay, and
3—Nanocomposite with 4 weight% of organoclay.

The comparison of the experimental data for the nanocomposites on the basis of the nylon-6, polypropylene and polystyrene gathered in Table 7 show that the heat of combustion, the smoke release and the amount of the carbon monoxide are almost constant at varying organoclay content. So we conclude that the source for the increased refractoriness of these materials is the stability of the solid phase and not the influence of the vapor phase. The data for the polystyrene with the 30% of the dekabrominediphenyloxide and Sb_2O_3 are given in Table 6 as the proof of the influence of the vapor phase of bromine. The incomplete combustion of the polymeric material in the latter case results in low value of the heat of the combustion and high quantity of the carbon monoxide released [79].

One should note that the mechanism for the increased fire resistance of the polymeric nanocomposites on the basis of the organoclays is not, in fact, clear at all. The formation of the barrier from the clay layers during the combustion at their collapse is supposed to be the main mechanism. That barrier slows down the combustion [80]. In chapter, we study the influence of the nanocomposite structure on the refractoriness. The layered structure of the nanocomposite expresses higher refractoriness comparing to that in intercalated nanocomposite, see Figure 17.

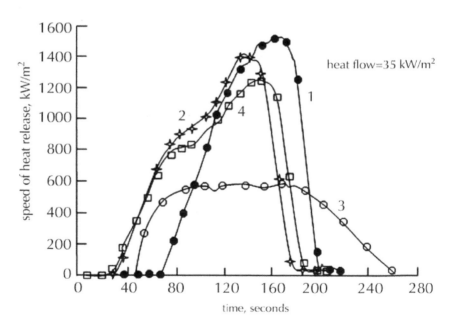

FIGURE 17   The speed of the heat release for:

1—Pure polystyrene (PS),
2—PS mixed with 3 weight% of Na⁺ MMT,
3—Intercalated/delaminated PS (3weight% 2C18-MMT) extruded at 170°C, and
4—Intercalated PS (3 weight% C14-FH) extruded at 170°C.

The data on the polymeric polystyrene–based nanocomposites presented in Figure 17 are for (1) initial ammoniumfluorine hectorite and (2) quaternary ammonium montmorillonite. The intercalated nanocomposite was produced in first case whereas the layered-intercalated nanocomposite was produced in the second one. But because the chemical nature and the morphology of the organoclay used was quite different it is very difficult to draw a unique conclusion about the flame resistance in polymeric nanocomposites produced. Nonetheless, one should point out that good results of the same quality were obtained for both layered and intercalated structures when studying the aliphatic groupings of the polyimide nanocomposites based on these clays. The better refractoriness is observed in case of polystyrene embedded in layered nanocomposite while intercalated polystyrene–based nanocomposite (with MMT) also exhibits increased refractoriness.

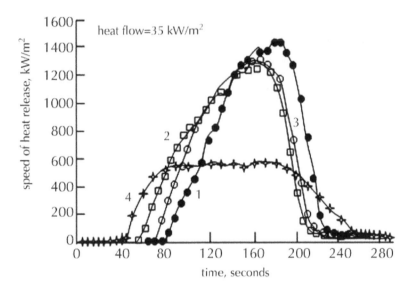

FIGURE 18   The speed of the heat release for:

1—Pure polysterene (PS),
2—Polystyrene with Na-MMT,
3—Intercalated PS with organomontmorillonite obtained in extruder at 185°C, and
4—Intercalated/layered PS with organoclay obtained in extruder at 170°C in nitrogen atmosphere or in vacuum.

As one can see the from the Figure 18, the speed of the heat release for the nanocomposite produced in nitrogen atmosphere at 170°C is much lower than for other samples. Probably, the reason for the low refractoriness of the nanocomposite produced in extruder without the vacuuming at 180°C is the influence of the high temperature and of the oxygen from the air what can lead to the destruction of the polymer in such conditions of the synthesis.

It is impossible to give an exact answer on the question about how the refractoriness of organoclay – based nanocomposites increases basing on only the upper experimental data but the obvious fact is that the increased thermal stability and refractoriness are due to the presence of the clays existing in the polymeric matrix as nanoparticles and playing the role of the heat isolators and elements preventing the flammable products of the decomposition to fly off.

There are still many problems unresolved in the field but indisputably polymeric nanocomposites will take the leading position in the chemistry of the advanced materials with high heat and flame resistance. Such materials can be used either as itself or in combination with other agents reducing the flammability of the substances.

The processes of the combustion are studied for the number of a polymeric nanocomposites based on the layered silicates such as nylon-6.6 with 5 weight% of Cloisite 15A–montmorillonite being modified with the dimethyldialkylammonium (alkyls studied C_{18}, C_{16}, C_{14}), maleinated polypropylene and polyethylene, both (1.5%) with 10 weight% Cloisite 15A. The general trend is two times reduction of the speed of the heat release. The decrease in the period of the flame induction is reported for all nanocomposites in comparison with the initial polymers [54].

The influence of the nanocomposite structure on its flammability is reflected in the Table 7. One can see that the least flammability is observed in delaminated nanocomposite based on the polystyrene whereas the flammability of the intercalated composite is much higher [54].

TABLE 7   Flammability of several polymers and composites

Sample	Coke remnant, weight%	Max speed of heat release, kW/m²	Average value of heat release, kW/m²	Average heat of combustion, MJ/kg	Specific smoke release, m²/kg	CO release, kg/kg
Nylon-6	1	1010	603	27	197	0.01
Nylon-6 + 2% of silicate (delaminated)	3	686	390	27	271	0.01
Nylon-6 + 5% of silicate (delaminated)	6	378	304	27	296	0.02
Nylon-12	0	1710	846	40	387	0.02
Nylon-12 + 2% of silicate (delaminated)	2	1060	719	40	435	0.02
Polystyrene	0	1562	803	29	1460	0.09
PS + 3% of silicate Na-MMT	3	1404	765	29	1840	0.09
PS + 3% of silicate C14-FH (intercalated)	4	1186	705	28	1790	0.09
PS + 3% of silicate 2C18-MMT (delaminated)	4	567	444	28	1730	0.08

TABLE 7 *(Continued)*

Polypropylene	0	1525	536	39	704	0.02
PP + 2% of silicate (intercalated)	3	450	322	40	1028	0.02

The optical properties of the nanocomposites are of much interest too. The same materials could be either transparent or opaque depending on certain conditions. For example in Figure 19 we see transparency, plot (a), and turbidity, plot (c), of the material in dependence of the frequency of the current applied.

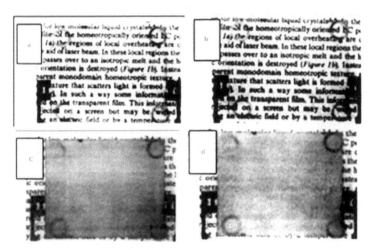

FIGURE 19 The optical properties of the clay – based nanocomposites in dependence of the applied electric current:

- (a) Low frequency, switched on,
- (b) Low frequency, switched off,
- (c) High frequency, switched on, and
- (d) High frequency, switched off.

The effect in Figure 19 is reversible and can be innumerately repeated. The transparent and opaque states exhibit the memory effect after the applied current switched off, plots (b) and (d) in Figure 19. The study of the intercalated nanocomposites based on the smectite clays reveals that the optical and elecrooptical properties depend on the degree of intercalation [81].

1.7 CONCLUSION

The quantity of the papers in the field of the nanocomposite polymeric materials has grown multiple times in recent years. The possibility to use almost all polymeric and polycondensated materials as a matrix is shown. The nanocomposites from various organoclays and polymers have been synthesized. Here is just a small part of the compounds for being the matrix referenced in literature: polyacrylate [83], polyamides [82,84,85], polybenzoxasene [86], polybutyleneterephtalate [11,82,87], polyimides [88], polycarbonate [89], polymethylmetacrylate [90], polypropylene [91,92], polystyrene [90], polysulphones [93], polyurethane [94], polybuthyleneterephtalate and polyethyleneterephtalate [10,65,68,79,95,99-107], polyethylene [96], epoxies [97].

The organo modified montmorillonite is of the special interest because it can be an element of the nanotechnology and it can also be a carrier of the nanostructure and of asymmetry of length and width in layered structures. The organic modification is being usually performed using the ion-inducing surface-active agents. The non-ionic hydrophobisation of the surface of the layered structures have been reported either. The general knowledge about the methods of the study is being formed and the understanding of the structure of the nanocomposite polymeric materials is becoming clear. Also, scientists come closer to the realizing of the relations between the deformational and strength properties and the specifics of the nanocomposite structure. The growth of researches and their direction into the nanoarea forecasts the fast broadening of the industrial involvement to the novel and attractive branch of the materials science.

REFERENCES

1.   Romanovsky, B.V. and Makshina, E. V. *Sorosovskii obrazovatelniu zhurnal* [in Russian], **8**(2), 50–55 (2013).
2.   Golovin, Yu. I. *Priroda* [in Russian], **1** 2009.
3.   Carter, L. W., Hendrics, J. G., and Bolley, D. S. United States Patent №2,531, 396 (1990).
4.   Blumstain, A. *Bull Chem Soc*, 899–905 (1991).
5.   Fujiwara, S. and Sakamoto, T. Japanese Application № 109, 998 (1996).
6.   Usuki, A., Kojima, Y., Kawasumi, M., Okada, A., Fukushima, Y., Kurauchi, T., and Kamigatio, O. *Journal of Appl polym science*, **55**, 119 (1995).
7.   Usuki, A., Koiwai, A., Kojima, Y., Kawasumi, M., Okada, A., Kurauchi, T., and Kamigaito, O. *Journal of Appl polym science*, **55**, 119 (1995)
8.   Okada, A. and Usuki, A. *Mater Sci Engng*, **3**, 109 (1995).
9.   Okada, A., Fukushima, Y., Kawasumi, M., Inagaki, S., Usuki, A., Sugiyama, S., Kurauchi, T., and Kamigaito, O. United States Patent №4,739,007 (1988).
10. Mikitaev, M. A., Lednev, O. B., Kaladjian, A. A., Beshtoev, B. Z., Bedanokov, A. Yu., and Mikitaev, A. K. Second International Conference, Nalchik (2005)
11. Chang, J. H., An, Y. U., Kim, S. J., and Im, S. *Polymer*, **44**, 5655–5661 (2003).
12. Mikitaev, A. K., Bedanokov, A. Y., Lednev, O. B., and Mikitaev, M.A. Polymer/silicate nanocomposites based on organomodified clays / Polymers, Polymer Blends, Polymer Composites and Filled Polymers. *Synthesis, Properties, Application*. Nova Science Publishers, New York (2006).
13. Delozier, D. M., Orwoll, R. A., Cahoon, J. F., Johnston, N. J., Smith, J. G., and Connell, J. W. *Polymer*, **43**, 813–822 (2002).
14. Kelly, P., Akelah, A., and Moet, A. *J. Mater. Sci.*, **29**, 2274–2280 (1994).
15. Chang, J. H., An, Y. U., Cho, D., and Giannelis, E. P. *Polymer*, **44**, 3715–3720 (2003).
16. Yano, K., Usuki, A., and Okada, A. *J Polym Sci, Part A: Polym Chem*, **35**, 2289 (1997).
17. Garcia-Martinez, J. M., Laguna, O., Areso, S., and Collar, E. P. *J Polym Sci, Part B: Polym Phys*, **38**, 1564 (2000).

18. Giannelis, E. P., Krishnamoorti, R., and Manias, E. *Advances in Polymer Science*, Vol.138, Springer-Verlag Berlin Heidelberg (1999).
19. Delozier, D. M., Orwoll, R. A., Cahoon, J. F., Ladislaw, J. S., Smith, J. G., and Connell, J. W. *Polymer*, 44, 2231–2241 (2003).
20. Shen, Y. H. *Chemosphere*, **44**, 989–995 (2001).
21. Giannelis, E. P. *Adv. Mater.*, **8** 29–35 (1996).
22. Dennis, H. R., Hunter, D. L., Chang, D., Kim, S., White, J. L., Cho, J. W., and Paul, D. R. *Polymer*, **42**, 9513–9522 (2001).
23. Kornmann, X., Lindberg, H., and Berglund, L. A. *Polymer* **42** 1303–1310 (2001).
24. Fornes, T. D. and Paul, D. R. Formation and properties of nylon 6 nanocomposites. São Carlos, *Polímeros*, **13**(4) (Oct/Dec, 2003).
25. Voulgaris, D. and Petridis, D. *Polymer*, **43**, 2213–2218 (2002).
26. Tyan, H. L., Liu, Y. C., and Wei, K. H. *Polymer*, **40**, 4877–4886 (1999).
27. Davis, C. H., Mathias, L. J., Gilman, J. W., Schiraldi, D. A., Shields, J. R., Trulove, P., Sutto, T. E., and Delong, H. C. *J Polym Sci, Part B: Polym Phys*, **40**, 2661 (2002).
28. Morgan, A. B. and Gilman, J. W. *J Appl Polym Sci*, **87**, 1329 (2003).
29. John, N. Hay and Steve, J. Shaw. Organic-inorganic hybridssthe best of both worlds? *Europhysics News*, **34**(3) (2003).
30. Chang, J. H., An, Y. U., and Sur, G. S. *J Polym Sci Part B: Polym Phys*, **41**, 94 (2003).
31. Chang, J. H., Park, D. K., and Ihn, K. J. *J Appl Polym Sci*, **84**, 2294 (2002).
32. Pinnavaia, T. J. *Science*, **220**, 365 (1983).
33. Messersmith, P. B. and Giannelis, E. P. *Chem Mater*, **5**, 1064 (1993).
34. Vaia, R. A., Ishii, H., and Giannelis, E. P. *Adv Mater*, **8**, 29 (1996).
35. Gilman, J. W. *Appl Clay Sci*, **15**, 31 (1999).
36. Fukushima, Y., Okada, A., Kawasumi, M., Kurauchi, T., and Kamigaito, O. *Clay Miner* **23**, 27 (1988).
37. Akelah, A. and Moet, A. *J Mater Sci*, **31**, 3589 (1996).
38. Vaia, R. A., Ishii, H., and Giannelis, E. P. *Adv Mater*, **8**, 29 (1996).
39. Vaia, R. A., Jandt, K. D., Kramer, E. J., and Giannelis, E. P. *Macromolecules*, **28**, 8080 (1995).
40. Greenland, D. G. *J Colloid Sci*, **18**, 647 (1963).
41. Chang, J. H. and Park, K. M. *Polym Engng Sci*, **41**, 2226 (2001).
42. Greenland, D. G. *J Colloid Sci*, **18**, 647 (1963).
43. Chang, J. H., Seo, B. S., and Hwang, D. H. *Polymer*, **43**, 2969 (2002).
44. Vaia, R. A., Jandt, K. D., Kramer, E. J., and Giannelis, E. P. *Macromolecules*, **28**, 8080 (1995).
45. Fukushima, Y., Okada, A., Kawasumi, M., Kurauchi, T., and Kamigaito, O. *Clay Miner*, **23**, 27(1988).
46. Chvalun, S. N. *Priroda* [in Russian], **7** (2000).
47. Brinker, C. J. and Scherer, G. W. *Sol-Gel Science*. Boston (1990).
48. Mascia, L. and Tang, T. *Polymer*, **39**, 3045 (1998).
49. Tamaki, R. and Chujo, Y. *Chem Mater*, **11**, 1719 (1999).
50. Serge Bourbigot, E. A. Investigation of Nanodispersion in Polystyrene–Montmorillonite Nanocomposites by Solid-State NMR. *Journal of Polymer Science: Part B: Polymer Physics*, **41**, 3188–3213 (2003).
51. Lednev, O. B., Kaladjian, A. A., Mikitaev, M. A., and Tlenkopatchev, M. A. New polybutylene terephtalate and organoclay nanocomposite materials. *Abstracts of the International Conference on Polymer materials*, México (2005).

52. Tretiakov, A. O. Oborudovanie I instrument dlia professionalov №02(37) [in Russian], (2003).
53. Sergeev, G. B. *Ros. Chem. J.* (The journal of the D. I. Mendeleev Russian chemical society) [in Russian], **46**(5) (2002).
54. Lomakin, S. M. and Zaikov, G. E. *Visokomol. Soed. B.* [in Russian], **47**(1) 104–120 (2005).
55. Xu, H., Kuo, S. W., Lee, J. S., and Chan, F. C. *Macromolecules*, **35**, 8788 (2002).
56. Haddad, T. S. and Lichtenhan, J. D. *Macromolecules*, **29**,7302 (1996).
57. Mather, P. T., Jeon, H. G., Romo-Uribe, A., Haddad, T. S., and Lichtenhan, J. D. *Macromolecules*, **29**, 7302 (1996).
58. Hsu, S. L. C. and Chang, K. C. *Polymer*, **43**, 4097 (2002).
59. Chang, J. H., Seo, B. S., and Hwang, D. H. *Polymer*, **43**, 2969 (2002).
60. Fornes, T. D., Yoon, P. J., Hunter, D. L., Keskkula, H., and Paul, D. R. *Polymer*, **43**, 5915 (2002).
61. Chang, J. H., Seo, B. S., and Hwang, D. H. *Polymer*, **43**, 2969 (2002).
62. Fornes, T. D., Yoon, P. J., Hunter, D. L., Keskkula, H., and Paul, D. R. *Polymer*, **43**, 5915 (2002).
63. Wen, J. and Wikes, G. L. *Chem Mater*, **8**, 1667 (1996).
64. Zhu, Z. K., Yang, Y., Yin, J., Wang, X., Ke, Y., and Qi, Z. *J Appl Polym Sci*, **3**, 2063 (1999).
65. Mikitaev, M. A., Lednev, O. B., Beshtoev, B. Z., Bedanokov, A. Yu, and Mikitaev, A. K. Second International conference *"Polymeric composite materials and covers"* [in Russian], Yaroslavl (may, 2005)
66. Fischer, H. R., Gielgens, L. H., and Koster, T. P. M. *Acta Polym*, **50**, 122 (1999).
67. Petrovic, X. S., Javni, L., Waddong, A., and Banhegyi, G. J. *J Appl Polym Sci*, **76**, 133 (2000).
68. Lednev, O. B., Beshtoev, B. Z., Bedanokov, A. Yu, Alarhanova, Z. Z., and Mikitaev, A. K. *Second International Conference* [in Russian], Nalchik (2005)
69. Chang, J. H., Kim, S. J., Joo, Y.L., and Im, S. *Polymer*, **45**, 919–926 (2004).
70. Lan, T. and Pinnavaia, T. *J. Chem Mater*, **6**, 2216 (1994).
71. Masenelli-Varlot, K., Reynaud, E., Vigier, G., and Varlet, J. *J Polym Sci Part B: Polym Phys*, **40**, 272 (2002).
72. Yano, K., Usuki, A., and Okada, A. *J Polym Sci Part A: Polym Chem*, **35**, 2289 (1997).
73. Shia, D., Hui, Y., Burnside, S. D., and Giannelis, E. P. *Polym Engng Sci*, **27**, 887 (1987).
74. Curtin, W. A. *J Am Ceram Soc*, **74**, 2837 (1991).
75. Chawla, K. K. *Composite materials science and engineering*. Springer, New York (1987).
76. Burnside, S. D. and Giannelis, E. P. *Chem. Mater.* 7, 4597 (1995).
77. Babrauskas, V. and Peacock, R. D. *Fire Safety Journal*, **18**, 225 (1992).
78. Gilman, J., Kashiwagi, T., Lomakin, S., Giannelis, E., Manias, E., Lichtenhan, J., and Jones, P., in. Fire Retardancy of Polymers: the Use of Intumescence. *The Royal Society of Chemistry*, Cambridge, 203–221 (1998).
79. Mikitaev, A. K., Kaladjian, A. A., Lednev, O. B., and Mikitaev, M. A. *Plastic masses* [in Russian], **12**, 45–50 (2004).
80. Gilman, J. and Morgan, A. 10th Annual BCC Conference (May 24–26, 1999).
81. John, N. Hay and Steve, J. Shaw. Organic-inorganic hybrids: the best of both worlds? *Europhysics News*, **34**(3) (2003).
82. Delozier, D. M., Orwoll, R. A., Cahoon, J. F., Johnston, N. J., Smith, J. G., and Connell, J. W. Polymer, **43**, 813–822 (2002).
83. Chen, Z., Huang, C., Liu, S., Zhang, Y., and Gong, K., J. *Apply Polym Sci*, **75**, 796–801 (2000).

84. Okado, A., Kawasumi, M., Kojima, Y., Kurauchi, T., and Kamigato, O. *Mater Res Soc Symp Proc*, **171**, 45 (1990).
85. Leszek, A. Utracki, Jorgen Lyngaae-Jorgensen. *Rheologica Acta*, 41, 394–407 (2002).
86. Wagener, R. and Reisinger, T. J. G. *Polymer*, **44**, 7513–7518 (2003).
87. Li, X., Kang, T., Cho, W. J., Lee, J. K., and Ha, C. S. *Macromol Rapid Commun*.
88. Tyan, H. L., Liu, Y. C., and Wei, K. H. *Polymer*, **40**, 4877–4886 (1999).
89. Vaia, R., Huang, X., Lewis, S., and Brittain, W. *Macromolecules*, **33**, 2000–2004 (2000).
90. Okamoto, M., Morita, S., Taguchi, H., Kim, Y., Kotaka, T., and Tateyama, H. *Polymer*, **41**, 3887–90 (2000).
91. Chow, W. S., Mohd Ishak, Z. A., Karger-Kocsis, J., Apostolov, A. A., and Ishiaku, U. S. *Polymer*, **44**, 7427–7440 (2003).
92. Antipov, E. M., Guseva, M. A., Gerasin, V. A., Korolev, Yu. M., Rebrov, A. V., Fisher, H. R., and Razumovskaya, I. V. *Visokomol soed. A*. [in Russian], **45**(11), 1885–1899 (2003).
93. Sur, G., Sun, H., Lyu, S., and Mark, *J. Polymer*, **42**, 9783–9789 (2001).
94. Wang, Z. and Pinnavaia, T. *Chem Mater*, **10**, 3769–3771 (1998).
95. Bedanokov, A. Yu., Beshtoev, B. Z. *Malij polimernij congress* [in Russian], Moscow (2005).
96. Antipov, E. M., Guseva, M. A., Gerasin, V. A., Korolev, Yu. M., Rebrov, A. V., Fisher, H. R., and Razumovskaya, I. V. *Visokomol. Soed. A*. [in Russian], **45**(11), 1874–1884 (2003).
97. Lan, T., Kaviartna, P., and Pinnavaia, T. *Proceedings of the ACS PMSE*, **71**, 527–528 (1994).
98. Kawasumi, et al. Nematic liquid crystal/clay mineral composites. *Science & Engineering C*, **6**, 135–143 (1998).
99. Lednev, O. B., Kaladjian, A. A., and Mikitaev, M. A. Second International Conference [in Russian], Nalchik (2005).
100. Mikitaev, A. K., Kaladjian, A. A., Lednev, O. B., Mikitaev, M. A., and Davidov, E. M. *Plastic masses* [in Russian], **4**, 26–31 (2005).
101. Eid, A., MikiTAEv, M. A., Bedanokov, A. Y., and Mikitaev, A. K. *Recycled Polyethylene Terephthalate/Organo-Montmorillanite Nanocomposites, Formation and Properties*. The first Afro-Asian Conference on Advanced Materials Science and Technology (AMSAT 06), Egypt (2006).
102. Mikitaev, A. K., Bedanokov, A. Y., Lednev, O. B., and Mikitaev, M. A. Polymer/silicate nanocomposites based on organomodified clays/ Polymers, Polymer Blends, Polymer Composites and Filled Polymers. *Synthesis, Properties, Application*. Nova Science Publishers, New York (2006).
103. Malamatov, A. H., Kozlov, G. V., and Mikitaev, M. A. Mechanismi uprochnenenia polimernih nanokompozitov [in Russian], RUChT, Moscow, p. 240 (2013)
104. Eid, A. Doctor Thesis, RUChT, Moscow, p. 121 (2013) [in Russian].
105. Lednev, O. B. Doctor Thesis, RUChT, Moscow, p. 12 (2013) [in Russian].
106. Malamatov, A. H. Professor Thesis, KBSU, Nalchik, p. 296 (2006) [in Russian].
107. Borisov, V. A., Bedanokov, A. Yu., Karmokov, A. M., Mikitaev, A. K., Mikitaev, M.,A., and Turaev, E. R. *Plastic masses*[in Russian], **5** (2013) .

CHAPTER 2

MULTI-SCALE MODELING AND SIMULATION OF DENDRITIC ARCHITECTURES: NEW HORIZONS

M. HASANZADEH and B. HADAVI MOGHADAM

CONTENTS

2.1 INTRODUCTION

The dendritic architectures, as highly branched and three-dimensional macromolecules that have unique chemical and physical properties, offer potential as the next great technological revolution. This review gives a brief introduction to some of the structural properties and application of dendritic polymer in various fields. The focus of the paper is a survey of multi-scale modeling and simulation techniques in hyperbranched polymer and dendrimers. Results of modeling and simulation calculations on dendritic architecture are reviewed.

The field of dendritic architectures, as a general class of macromolecules, has found widespread interest in the past decades. Much has been achieved in the preparation of three-dimensional structures such as comb- and star-shaped polymers and dendrimers. These materials have comparable physical and chemical properties to their linear analogous that make them very attractive for numerous applications [1-4].

Intensive studies in the area of dendritic macromolecules, which include applied research and are generally interdisciplinary, have created a need for a more systematic approach to dendritic architectures development that employs a multi-scale modeling and simulation approach. A possible way is to determine the atomic-scale characteristics of dendritic molecules using computer simulation and computational approaches. Computer simulation, as a powerful and modern tool for solving scientific problems, can be performed for dendritic architectures without synthesizing them. Computer simulation not only used to reproduce experiment to elucidate the invisible microscopic details and further explain experiments, but also can be used as a useful predictive tool. Currently, Monte Carlo, Brownian dynamics and molecular dynamics are the most widely used simulation methods for molecular systems [5].

The objective of this chapter is to address recent advances in molecular simulation methodologies and computational power. In this paper, we will first briefly review the structure, properties and application of dendritic macromolecules in various fields. Next, molecular simulation techniques in hyperbranched polymer and dendrimers will be reviewed. Lastly, we will survey the most characteristic and important recent examples in molecular simulation of dendritic architectures.

2.2 DENDRITIC ARCHITECTURES

2.2.1 BASIC PRINCIPLES

Dendritic architectures are highly branched polymers with tree like branching having an overall spherical or ellipsoidal shape and are known as additives having peripheral functional groups. These macromolecules consist of three subsets namely dendrimers, dendrigraft polymers and hyperbranched polymers (Figure 1).

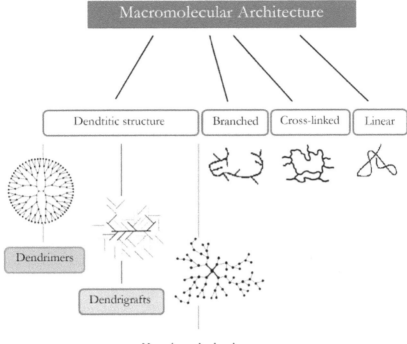

FIGURE 1   Classification of macromolecular architecture.

Dendrimers resemble star polymers except that each leg of the star exhibit repetitive branching in the manner of a tree. There are two general routes for the synthesis of the dendrimers: divergent methodology and convergent methodology (Figure 2). Dendrigraft (arborescent) polymers are prepared by linking macromolecular building blocks. In dendrigraft polymers branching sites are usually distributed randomly along the dendritic chains of the dendrigraft interior whereas in dendrimers the branching sites conducing to the next generation occur only at the chain end. Hyperbranched polymers (HBPs) are highly branched, polydisperse and three-dimensional macromolecules; and it is synthesized from a multifunctional monomer to produce a molecule with dendritic structure [6-16]. Table 1 shows typical characteristics of dendritic architectures.

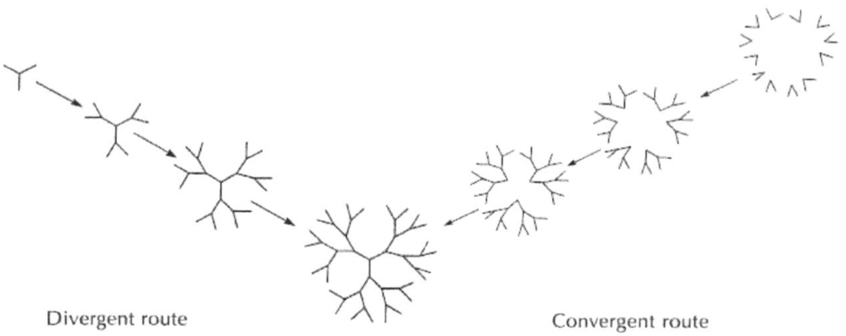

Divergent route   Convergent route

FIGURE 2 Two general routes for synthesis of dendrimers.

The dendrimers are well-defined and need a stepwise route to construct the perfectly symmetrical structure. Hence, synthesis of dendrimers is time-consuming and expensive procedures. Although, hyperbranched polymers are irregularly shaped and not perfectly symmetrical like dendrimers, hyperbranched polymers rapidly prepared and generally synthesized by one-step process via polyaddition, polycondensation, radical polymerization, and so on, of ABx (mostly AB2, equal reactivity of all Bs) type monomers and no purification steps are needed for their preparation (Figure 3). Therefore HBPs are attractive materials for industrial applications due to their simple production process [6-9, 17-25]. In general, according to molecular structures and properties, hyperbranched polymers represent a transition between linear polymers and perfect dendrimers [18]. The comparison of hyperbranched polymers with their linear analogues indicated that HBPs have remarkable properties, such as low melt and solution viscosity, low chain entanglement, and high solubility, as a result of the large amount of functional end groups and globular structure [16-22].

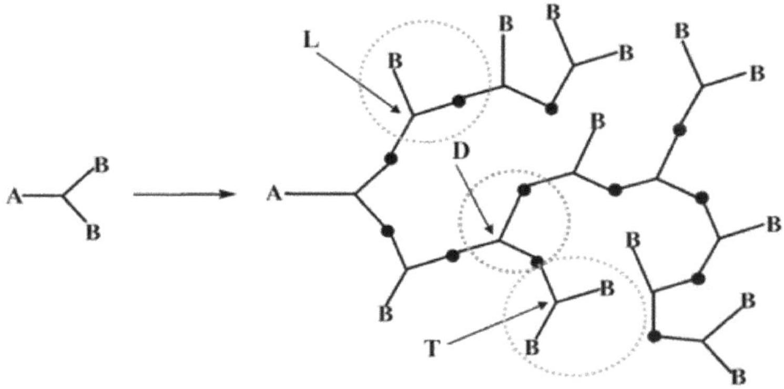

FIGURE 3 Schematic representation of a hyperbranched polymer construction and its structural units include terminal (T), dendritic (D), and linear (L) units.

TABLE 1   Comparison of different dendritic polymers

Properties	Dendritic architectures		
	Dendrimers	Dendrigraft	Hyperbranched
Terminal units	Small	Linear chains	Small
Molecular mass distribution	Narrow	Narrow	Broad
Synthetic steps	4-20	2-5	1
Purification steps	4-20	2-5	0
Cost	Very high	Moderate	Low

2.3   GENERAL STRUCTURAL CONSIDERATIONS

2.3.1   DEGREE OF BRANCHING

The different structural parameters such as degree of polymerization, degree of branching and Wiener index can be used to characterize the topologies of hyperbranched polymers. The degree of branching (DB) is defined as follows:

$$DB = \frac{2D}{(2D+L)} \tag{1}$$

Where D is the number of dendritic units and L is the number of linear units. This value varies from 0 for linear polymers to 1 for dendrimers or fully branched hyperbranched polymers [26-28].

2.3.2   WIENER INDEX

In addition to the degree of branching, the Wiener index is also used to distinguish polymers of different topologies and defined as:

$$W = \frac{1}{2}\sum_{j=1}^{N_s}\sum_{i=1}^{N_s} d_{ij} \tag{2}$$

Where N_s is the number of beads per molecule and d_{ij} is the number of bonds separating site i and j of the molecule. This parameter only describes the connectivity and is not a direct measure of the size of the molecules. Larger Wiener index numbers indicate higher numbers of bonds separating beads in molecules and hence more open structures of polymer molecules [28]. Table 2 shows the DB of polymers with different architecture and the same degree of polymerization.

TABLE 2 Degree of branching for different polymer architectures of the same molecular weight (white beads representing linear units and gray beads representing branching units).

Polymer type	Polymers architecture	Degree of Branching
Linear polymer		DB=0
Hyper-branched polymer		0<DB<1
Dendrimers		
		DB=1
Fully branched hyperbranched polymer		

2.3.3   RADIUS OF GYRATION

The long chain branching on the polymer s can have a major effect on the rheological properties. The comparison of a long branching polymer with a linear polymer with the same molecular weight shows a drastic decrease in the mean-square dimensions of the polymer in solution. The radius of gyration, which is the trace of the tensor of gyration, can describe the size of a polymer molecule. The mean square radius of gyration tensor of molecules can be calculated according to the formula:

$$\left\langle R_g R_g \right\rangle \equiv \left\langle \frac{\sum\limits_{\alpha=1}^{N_s} m_\alpha \left(r_\alpha - r_{CM}\right)\left(r_\alpha - r_{CM}\right)}{\sum\limits_{\alpha=1}^{N_s} m_\alpha} \right\rangle \tag{3}$$

Where r_α is the position of site α, r_{CM} is the position of the molecular center of mass, m_α is the mass of site α and the angle brackets denote an ensemble average. The value of the squared radius of gyration is defined as the trace of the tensor

$$R_g^2 = Tr\left(\left\langle R_g R_g \right\rangle\right)$$

By studying the tensor of gyration, the shape of hyperbranched polymers can be investigated. Polymers with higher degree of polymerization normally have larger radius of gyration. Furthermore for a given value of molecular weight, the value of the radius of gyration for different branched polymers can vary depending on the branching topology [28].

2.3.4   DISTRIBUTION OF MASS

The radial distribution function is a useful tool to describe the structure of dendritic structure. The distribution of sites from the molecular center of mass is given as:

$$g_{CM} = \frac{\left\{\sum\limits_{i=1}^{N}\sum\limits_{\alpha=1}^{N_s} \delta\left(\left|r - (r_{i\alpha} - r_{CM})\right|\right)\right\}}{N} \tag{4}$$

Where N is the total number of molecules, r_{CM} is the position of the center of mass and α runs over all other sites belonging to the same molecule. Similarly, the distribution from the central site (the core) can be defined as:

$$g_{core}(r) = \frac{\left\{\sum\limits_{i=1}^{N}\sum\limits_{\alpha=1}^{N_s} \delta\left(\left|r - (r_{i\alpha} - r_{i1})\right|\right)\right\}}{N} \tag{5}$$

Where r_{i1} is the position of the core. Another useful function for characterization of internal structure and spatial ordering of sites composing the materials is the atomic radial distribution.

$$g_A(r) = \frac{\left\{ \dfrac{1}{2} \sum_{i=1}^{N_{total}} \sum_{j \neq i}^{N_{total}} \delta\left(\left|r - r_{ij}\right|\right) \right\}}{4\pi r^2 N_{total}\rho} \tag{6}$$

Where r_{ij} is the distance between the sites i and j, $N_{total} = NNs$ is the total number of sites in the studied system, and ρ is the density [28].

2.4 APPLICATIONS

A number of excellent reviews have been published on synthesis, functionalization and applications of dendritic polymer [7,17,18,29]. Many application fields based on dendritic polymer have been steadily extended especially in recent years. A schematic diagram illustrating perspective application areas is summarized in Figure 4. Although some of these applications have not reached their industry level, their promising potential is believed to be attracting attentions and investments from academia, governments, and industry all over the world.

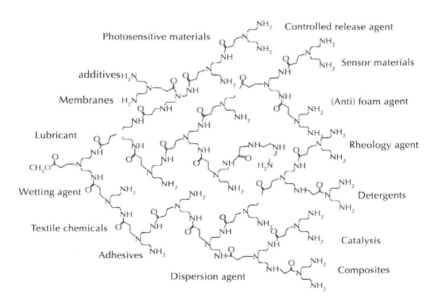

FIGURE 4   Potential application of dendritic polymers.

Dendritic polymers and their substitutes can be used as nanomaterials for host-guest encapsulation [30], fabrication of organic-inorganic hybrids [31] and nanoreactors [32]. Moreover, due to the low cost and well-defined architecture with multifunctional terminal groups, these types of polymers have recently attracted special interest in biomaterials application, as biocarriers and biodegradable materials [33]. Furthermore, based on the unique chemical and physical properties, hyperbranched polymer have been used as cross-linking or adhesive agents [34], dye-receptive additives [35], rheology modifiers or blend components [36], tougheners for thermosets [37], and so on. Table 3 shows dendritic architectures applications in various fields and their main characteristics.

In the case of textile chemistry, studies have demonstrated that the dyeability of textile fabrics, such as cotton [38-41], poly (ethylene terephthalate) (PET) [3,42,43], polyamide-6 [44], and polypropylene (PP) [35], were significantly improved by hyperbranched polymer. In the most recent investigation in this filed, the dyeability of modified PET fabrics by amine-terminated HBP was investigated by Hasanzadeh et al. [3,42]. They reported that the dye uptake of HBP-treated PET fabrics is significantly greater than that of untreated PET ones due to the presence of terminal primary amino groups in the molecular structure of the HBP, that will protonate in the liquid phase and give rise to positive charge at lower pH values.

TABLE 3   Dendritic architectures applications in various fields and its main characteristics

Application	Main Characteristics	Ref.
	Modification of PET fabrics for improving dyeability of acid dyes	[3]
	Modification of fiber grade PET for dyeing of disperse dyes	[43]
	Modification PA6 fabrics for studying dyeability with acid dyes	[44]
Textile dyeing	Dyeing of modified polypropylene fibers	[35]
	Reactive dyeing on silk fabric	[45]
	Antibacterial activity and anti-ultraviolet properties	[46]
	Modification of cotton for improving dyeability	[38-41]
	Modification of cotton for improving dyeability of reactive dyes	[47]
Drug delivery	Synthesis of water-soluble and degradable hyperbranched polyesters for drug delivery	[48]
DNA delivery	Degradability, cytotoxicity, and in vitro DNA transfection efficiency of the second type of hyperbranched poly(amino ester)s with different terminal amine groups	[49]

Gene carrier	Degradability, great ability to form complexes with DNA and suitable physicochemical properties of poly(ester amine)	[50]
Gene delivery	Biodegradability, very low toxicity and the ability to transfect cells	[51]
Metal ion extractant	Synthesis and application of new nitrogen centered hyperbranched polyesters	[52]
Gas separation	Synthesis and application of a series of wholly aromatic hyperbranched polyimides for gas separation application	[53]
Extractive distillation and solvent extraction	Cost-savings of extractive distillation using commercially available hyperbranched polymers	[54]
Fuel cells	Ionic conductivity, thermal properties, and fuel cell performance of the cross-linked membrane	[55]
Coating materials	Synthesis of fluorinated hyperbranched polymers holding different surface free energies and their applications as coating materials to afford highly hydrophobic and/or oleophobic cotton fabrics by solution-immersion method	[56]
Cationic UV curing	Effect of the hyperbranched polyol as flexibilizer and chain-transfer agent on cationic UV curing	[57]

2.5   MOLECULAR SIMULATION METHODS

According to the literature, the molecular simulation methods for nanosystems can be mainly classified into two categories:

i)   Atomistic modeling, and

ii) Continuum mechanics approaches [58-70].

In general, the atomistic modeling technique includes three important categories, namely the molecular dynamics (MD), Monte Carlo (MC), and *ab initio* approaches. In general, *ab initio* approaches are constructed on the basis of an accurate solution of the Schrödinger equation to extract the locations of each atom. Meanwhile, the main objective of MD and MC simulation is to solve the governing equations of particle dynamics based on the second Newton's law. In addition to these methods, tight bonding molecular dynamics (TBMD), local density approximation (LDA), and density functional theory (DFT) have also been proposed [71], but they are often computationally expensive. Although atomistic modeling technique provides a valuable insight into complex structures, due to its huge computational effort especially for large-scale systems, its application is limited to the systems with small number of atoms. Therefore, the alternative continuum and multiscale models were proposed for larger systems or larger time.

2.5.1 MONTE CARLO SIMULATION

The Monte Carlo method is a computerized mathematical technique that relies on repeated random sampling to obtain numerical results. This technique generates large numbers of configurations or microstates of equilibrated systems by stepping from one microstate to the next in a particular statistical ensemble. The advantage of the Monte Carlo simulation technique is that the probability distributions within the model can be easily and flexibly used, without the need to approximate them. The underlying matrix or trial move which must satisfy the principle of microscopic reversibility and time saving as only the potential energy is required, are some advantage of Monte Carlo simulation [72-73].

2.5.2 MOLECULAR DYNAMICS SIMULATION

The MD simulation is one of the most important techniques to study the macroscopic behavior of systems by following the evolution of the system at the molecular scale. This method is in many respects very similar to real experiments. As mentioned above, this technique deals with the case of Newton's equations of classical mechanics. Coupled Newton's equations of motion, which describe the positions and momenta, are solved for a large number of particles in an isolated cluster or in the bulk using periodic boundary conditions. The equations of motion for these particles which interact with each other via intra- and inter-molecular potentials can be solved accurately using various numerical integration methods such as the common predictor-corrector or Verlet methods. Molecular dynamics efficiently evaluates different configurational properties and dynamic quantities which cannot generally be obtained by Monte Carlo [74]. These configurations provide information on the types of structures and also completely describe the detailed dynamics of structural change and energy flow within the classical model. More details can be found in [75,76].

2.5.3 INTERMOLECULAR INTERACTION

It is believed that the particles interact with each other and, moreover, may be subject to external influence. Interatomic forces are represented in the form of the classical potential force (the gradient of the potential energy of the system). The interaction between atoms is described by means of van der Waals forces (intermolecular forces), mathematically expressed by the Lennard–Jones (LJ) potential [75]:

$$U_{ij}^{LJ} = 4\varepsilon \left[\left(\frac{\sigma}{r_{ij}} \right)^{12} - \left(\frac{\sigma}{r_{ij}} \right)^{6} \right] \tag{7}$$

Where r_{ij} is the distance between ith and jth particles normalized by typical length σ, and ε is the potential well depth, which governs the strength of interaction. It should be noted that the second term corresponds to the attractive force, which is caused by instantaneous induced van der Waals dipole-dipole attraction between atoms, and the first term corresponds to the Pauli repulsion. Generally, in order to shorten the computer time, the pair interactions beyond the distance of r_c are neglected. Therefore the LJ potential actually used in simulations is a truncated version defined as: (typically $r_c = 2.5\sigma$)

$$U_{ij}^{LJ} = 4\varepsilon\left[\left(\frac{\sigma}{r_{ij}}\right)^{12} - \left(\frac{\sigma}{r_{ij}}\right)^{6}\right], \quad for \quad r_{ij} \le r_c$$

$$U_{ij}^{LJ} = 0 \qquad\qquad\qquad for \quad r_{ij} \rangle r_c$$

(8)

This potential gives a reasonable representation of intermolecular interactions in noble gases, such as Ar and Kr, and systems composed of almost spherical molecules, such as methane. To get an idea of the magnitude, for Ar values are $\dfrac{\varepsilon}{k_B} = 120K$ and $\sigma = 0.34nm$.

Figure 6 shows the chart of the potential energy of intermolecular interaction.

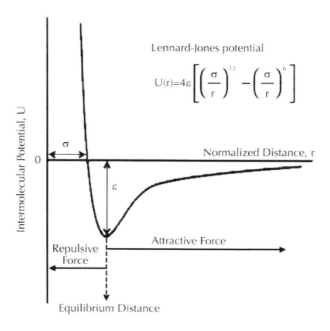

FIGURE 5   Potential energy of intermolecular interaction.

2.6 MOLECULAR SIMULATION IN DENDRITIC ARCHITECTURES

2.6.1 BACKGROUND

According to the literature, most experimental studies have been provided vital information on the large-scale interactions between dendritic molecules and other molecules, but many atomic-level questions and unsolved problems, which cannot be answered by experiments, are also pointed out. Therefore, theoretical and computational approaches in determining the atomic-scale characteristics of dendritic molecules have attracted a great interest. A tremendous amount of theoretical research has been applied to investigate the structure and dynamics of dendritic polymers. Recent advances in simulation methods and computational power have made it possible to study the physicochemical properties of dendritic polymers and their interaction with other molecules.

2.6.2 HYPERBRANCHED POLYMER

In 1998 the first numerical studies of hyperbranched polymer was reported by Aerts [77] in which configurations of hyperbranched polymers were modeled using the bead model of Lescanec and Muthukumar [78] and intrinsic viscosities were calculated. The application of the algorithm with no configurational relaxation and the questionable deduction of the intrinsic viscosity from the radius of gyration of branched structures were limitations of this work. In the study on structural properties of hyperbranched polymer in the melt under shear using a coarse-grained model and non-equilibrium molecular dynamics (NEMD) techniques by Le et al. [79], it was demonstrated that most of the microscopic structural properties have significant changes induced by the shear flow, which depend on the size and geometry of the molecules. Analysis of the tensor of gyration shows that hyperbranched polymer molecules have a prolate ellipsoid shape under shear which is slightly flatter than the ellipsoid shape of dendrimers. They also found that the conformational behavior of large hyperbranched polymers is similar to that of linear polymers whereas for small hyperbranched polymers, the behavior is similar to that of dendrimers. Moreover, the distribution of terminal beads was investigated and the results show that simulated hyperbranched polymers have similar distribution of mass with terminal groups existing everywhere inside the molecules [79]. Lyulin et al. [80] employed Brownian dynamics simulations of hyperbranched polymers with different degrees of branching (DB) under shear with excluded-volume and hydrodynamic interactions to obtain the intrinsic viscosity data. The results show that hyperbranched polymers with low values of DB reveal sparse structures while those with high values of DB possess very compact structures. Moreover, as the molecular weight of highly branched structures increases, the zero shear rate intrinsic viscosity reaches a maximum and begins to fall similar to the intrinsic viscosity behavior of dendrimers [80]. In another study, theoretical models for the structures of randomly hyperbranched polymers in solution were derived by Konkolewicz et al. [81]. Their models are based on the random assembly of various simple units which can be monomers, linear chains, or larger branched species. They presented a comparison of Monte

Carlo and molecular dynamics simulation with theoretical model conformation results including the radii of gyration and full density profiles for randomly hyperbranched polymers in solution [81].

2.6.3   DENDRIMERS

de Gennes and Hervet [82], for the first time used a mean field approach to determine the scaling properties of dendrimers. They reported that the size of dendrimers changes with the number of monomers as $R_g \propto N^{1/5}$, where R_g is the mean molecular radius of gyration. Maiti et al. [83] performed the first atomistic based MD simulations of G4 to G6 polyamidoamine (PAMAM) dendrimers in explicit water. The radius of gyration calculated from the simulation is roughly in agreement with experimental findings. Also, they found that the extent of ionization significantly affected the R_g of the dendrimer [83]. Cagin et al. [84-85] used the Continuous Configurational Boltzmann Biased (CCBB) direct Monte Carlo method in building the 3D molecular representations of dendrimers. The energetic and structural properties of dendrimers were also studied using molecular dynamics after annealing these molecular representations. Atomistic simulation using a self-avoiding walk and also molecular dynamics simulation were performed by some researchers [78]. Lue et al. [86] studied the behavior of dendrimers in dilute and concentrated solution using Monte Carlo simulation. The equation of state and the second virial coefficients was calculated by them. By comparison with linear chain systems, it is found that for low concentration, the second virial coefficients and the pressure of dendrimers were smaller than for linear chain molecules. Other series of equilibrium molecular simulation studies analyzed and reported the properties of isolated dendrimers [87]. According to the literature, most theoretical and computational studies of dendrimers reported the properties of isolated molecules or molecules in dilute solution. As a first atomistic simulation of dendrimers in the melt, Zacharopoulos et al. [88] reported the morphology of poly(propyleneimine) dendrimers. They found that the redial density profile decrease with the distance from the center of mass. Moreover, the radius of gyration was reported to depend on the number of beads to the power of ~ 0.29. More information on recent advances in simulation of dendritic polymers is summarized in Table 4.

TABLE 4   Molecular simulation of dendritic architectures in the literature

Dendritic polymer	Method	Focus of the research	Ref.
Dendrimers	Atomistic based molecular dynamics	Properties for generations up to the limiting growth size	[83]
	Molecular dynamics	Generation 2 through 6 at several pH values	[89]
	Molecular dynamics	The static properties of the PAMAM dendrimers in aqueous solution	[90]
	Continuous Configurational Boltzmann Biased (CCBB) direct Monte Carlo	Building the 3D molecular representations of dendrimers and study the energetic and structural properties of dendrimers	[80-81]
	Rouze-Zimm hydrodynamic	Dynamic properties of dendrimers in dilute solution	[91]
	Monte Carlo	The behavior of dendrimers in dilute solution	[86]
	Atomistic simulation	Morphology of poly(propyleneimine) dendrimers in melt	[88]
	Metropolis Monte Carlo algorithm	Calculating the quantitative structure–affinity relationship in a dendrimer–drug system	[92]
Hyperbranched polymers	Reverse Monte Carlo	Generation of randomly branched polymers with different architectures and sizes in solution.	[81]
	Non-equilibrium molecular dynamics	Structural properties of hyperbranched polymers in the melt under shear	[79]
	Monte Carlo	Copolymerization of a self-condensing vinyl monomer and a conventional vinyl monomer in the presence of a multifunctional initiator at equal rate constants	[93]
	Brownian dynamics	Simulations of hyperbranched polymers with different degrees of branching	[80]
	Brownian dynamics	Simulations of hyperbranched polymers up to the sixth generation under elongational flow	[94]
	Brownian dynamics	The structure and transport properties of dendritic polymers in dilute solution	[95]
	Brownian dynamics	Simulations of complexes formed by hyperbranched polymers with linear polyelectrolytes under steady shear flow	[96]
	Monte Carlo and molecular dynamics simulation	Testing theoretical models for randomly hyperbranched polymers in solution.	[97]

2.7 CONCLUSION

The dendritic architectures described in this review are highly branched and three-dimensional macromolecules that have new and unusual properties. Compared to their linear analogues, dendritic polymers are expected to have remarkable properties, such as lower melt and solution viscosity, lower chain entanglement, and higher solubility, as a result of the large amount of functional end groups and globular structure. In recent years, great efforts have been devoted to dendritic macromolecules. Although tremendous progress has been made in synthesis and characterization of dendritic polymers as well as their application during past decade, we have only vague answer concerning the atomic-scale characteristics of dendritic molecules. Much useful information about the atomic-scale characteristics of dendritic polymers has been gained in recent years using molecular simulation techniques, such as Monte Carlo, Brownian dynamics simulations and molecular dynamics. Molecular simulations provide atomic-scale insights into structure and properties of dendritic polymers and their interactions with other molecules. This information from simulations has successfully matched experimentally measured properties, and can help in the rational design of dendritic architectures for application in many areas.

REFERENCES

1. Yan, D., Gao, C., and Frey, H., *Hyperbranched Polymers Synthesis, Properties, and Applications*, John Wiley & Sons, New Jersey (2011).
2. Frechet, J. M. J. and Tomalia, D. A. *Dendrimers and other dendritic polymers*, John Wiley & Sons, UK (2011).
3. Hasanzadeh, M., Moieni, T, and Hadavi Moghadam, B. 'Synthesis and characterization of an amine terminated AB_2-type hyperbranched polymer and its application in dyeing of poly (ethylene terephthalate) fabric with acid dye', *Advances in Polymer Technology*, **32**, 792–799 (2013).
4. Mishra, M. K. and Kobayashi, S. Star and Hyperbranched Polymers. *Marcel Dekker*, New York (1999).
5. Gates, T. S., Odegard, G. M., Frankland, S. J. V., and Clancy, T. C. Computational materials: Multi-scale modeling and simulation of nanostructured materials. *Composites Science and Technology*, **65** 2416–2434 (2005).
6. Jikei, M., and Kakimoto M. Hyperbranched polymers: a promising new class of materials. Progress in Polymer Science, **26**, 1233–1285 (2001).
7. Gao, C, Yan, D. Hyperbranched polymers: from synthesis to applications. *Progress in Polymer Science*, **29**, 183–275 (2004).
8. Voit, B. I and Lederer, A. Hyperbranched and highly branched polymer architectures synthetic strategies and major characterization aspects. *Chem. Rev.*, **109**, 5924–5973 (2009).
9. Kumar, A. and Meijer, E. W. Novel hyperbranched polymer based on urea linkages. *Chem. Commun.*, 1629–1630 (1998).
10. Grabchev, I., Petkov, C., and Bojinov, V. Infrared spectral characterization of poly (amidoamine) dendrimers peripherally modified with 1,8-naphthalimides. *Dyes and Pigments*, **62**, 229–234 (2004).
11. QingHua, C, RongGuo, C, Li-Ren, X, Qing-Rong, Q, and Wen-Gong, Z. Hyperbranched poly (amide-ester) mildly synthesized and its characterization. Chinese J. *Struct. Chem.*, **27**, 877–883 (2008).

12. Kou, Y, Wan, A. Tong S. Wang, L, and Tang, J. Preparation, characterization and modification of hyperbranched polyester-amide with core molecules. *Reactive & Functional Polymers*, **67**, 955–965 (2007).
13. Schmaljohann, D. P. Pötschke. P. Hässler. R. Voit, B. I., Froehling. P. E. Mostert, B, Loontjens, and J. A. Blends of amphiphilic, hyperbranched polyesters and different polyolefins. *Macromolecules*, **32**, 6333–6339 (1999).
14. Kim, Y. H. Hyperbranched polymers 10 years after. *Journal of Polymer Science: Part A: Polymer Chemistry*, **36**, 1685–1698 (1998).
15. Liu, G. and Zhao, M. Non-isothermal crystallization kinetics of AB3 hyper-branched polymer/polypropylene blends. *Iranian Polymer Journal*, **18**, 329–338 (2009).
16. Inoue, K. Functional dendrimers, hyperbranched and star polymers, Progress in polymer science 2000; 25: 453-571.
17. Seiler, M. Hyperbranched polymers: Phase behavior and new applications in the field of chemical engineering. *Fluid Phase Equilibria*, **241**, 155–174 (2006).
18. Yates, C. R. and Hayes, W. Synthesis and applications of hyperbranched polymers. *European Polymer Journal*, **40**, 1257–1281 (2004).
19. Voit, B. New developments in hyperbranched polymers. *Journal of Polymer Science: Part A: Polymer Chemistry*, **38**, 2505–2525 (2000).
20. Nasar, A. S., Jikei, M., and Kakimoto, M. Synthesis and properties of polyurethane elastomers crosslinked with amine-terminated AB2-type hyperbranched polyamides. *European Polymer Journal*, **39**, 1201-1208 (2003).
21. Froehling, P. E. Dendrimers and dyes-a review. *Dyes and pigments*, **48**, 187–195 (2001).
22. Jikei, M., Fujii. K., and Kakimoto, M. Synthesis and characterization of hyperbranched aromatic polyamide copolymers prepared from AB_2 and AB monomers. *Macromol. Symp.*, 2003, **199**, 223–232 (2001).
23. Radke, W, Litvinenko, G, and Müller A. H. E. Effect of core-forming molecules on molecular weight distribution and degree of branching in the synthesis of hyperbranched polymers. *Macromolecules*, **31**, 239–248 (1998).
24. Maier, G, Zech, C. Voit, B., and Komber, H. An approach to hyperbranched polymers with a degree of branching of 100%. Macromol. *Chem. Phys.*, **199**, 2655–2664 (1998).
25. Voit, B, Beyerlein, D. Eichhorn, K. Grundke, K. Schmaljohann, D. and Loontjens, T. Functional hyper-branched polyesters for application in blends, coations, and thin films. *Chem. Eng. Technol*, **25**, 704–707 (2002).
26. Frey, H. and Hölter, D., Degree of branching in hyperbranched polymers. 3 Copolymerization of ABm-monomers with AB and ABn-monomers, *Acta Polym.*, **50**, 67–76 (1999).
27. Burgath, D. A. Frey, H. Degree of branching in hyperbranched polymers, *Acta Polymerica*, **48**, 30–35 (1997).
28. Le, Tu. C., Todd, B. D., Daivis, P. J., and Uhlherr, A. Structural properties of hyperbranched polymers in the melt under shear via nonequilibrium molecular dynamics simulation. *The Journal of Chemical Physics*, **130**, 074901 (2009).
29. Mohammad, Haj-Abed, *Engineering of Hyperbranched Polyethylene and its Future Applications*, MSc thesis, McMaster University, (2008).
30. Stiriba, S. E., Kautz, H., and Frey, H. Hyperbranched molecular nanocapsules: comparison of the hyperbranched architecture with the perfect linear analogue, *Journal of the American Chemical*, (2002)
31. Zou, J., Zhao, Y., Shi, W., X. Shen, and Nie, K. Preparation and characters of hyperbranched polyester-based organic-inorganic hybrid material compared with linear polyester, *Polymers for Advanced Technologies*, **16**, 55–60, (2005).

32. Zhu, L. Shi, Y. Tu, C. Wang, R. Pang, Y. Qiu F. Zhu, X. Yan, D. He, L. Jin, C., and Zhu, B. Construction and application of a pH-sensitive nanoreactor via a double-hydrophilic multiarm hyperbranched polymer, *Langmuir*, **26**(11), 8875– 81 (2010).

33. Zhou, Y. W., Huang, J. Liu, Zhu, X., and D. Yan. Self-Assembly of Hyperbranched Polymers and Its Biomedical Applications, *Advanced Materials*, **22**, 4567–4590 (2010)

34. Dodiuk-Kenig, H. and Lizenboim, K. The effect of hyper-branched polymers on the properties of dental composites and adhesives. *Journal of adhesion* (2004).

35. Burkinshaw, S. M., Froehling, P. E., and Mignanelli, M. The effect of hyperbranched polymers on the dyeing of polypropylene fibres, *Dyes and Pigments*.

36. Mulkern, T. J., Tan, N. C. Processing and characterization of reactive polystyrene/hyperbranched polyester blends, *Polymer* (2000)

37. Boogh, L, Pettersson, B. and Månson, J. A. E. Dendritic hyperbranched polymers as tougheners for epoxy resins, Polymer (1999).

38. Zhang, F., Chen, Y, Lin, H., and Lu. Y. Synthesis of an amino-terminated hyperbranched polymer and its application in reactive dyeing on cotton as a salt-free dyeing auxiliary. *Coloration Technology*, **123**, 351–357 (2007)

39. Zhang, F., Chen, Y., Lin H, Wang. H., and Zhao, B. HBP-NH2 grafted cotton fiber: Preparation and salt-free dyeing properties. *Carbohydrate Polymers*, **74**, 250–256 (2008).

40. Zhang, F, Chen, Y. Y., Lin, H., and Zhang D. S. Performance of cotton fabric treated with an amino-terminated hyperbranched polymer. *Fibers and Polymers*, **9**, 515–520 (2008).

41. Zhang, F, Chen, Y, Ling, H, and Zhang, D. Synthesis of HBP-HTC and its application to cotton fabric as an antimicrobial auxiliary. *Fibers and Polymers*, **10**, 141–147 (2009).

42. Hasanzadeh, M., Moieni, T., and Hadavi Moghadam, B., 'Modification of PET fabrics by hyperbranched polymer: A comparative study of artificial neural networks (ANN) and statistical approach', *Journal of Polymer Engineering*, 00, 00 (2013).

43. Khatibzadeh, M, Mohseni, M. and Moradian, S. Compounding fibre grade polyethylene terephthalate with a hyperbranched additive and studying its dyeability with a disperse dye. *Coloration Technology*, **126**, 269–274 (2010).

44. R. Mahmoodi, T. Dodel, T. Moieni, and M. Hasanzadeh, "Synthesis and Application of Amine-Terminated AB2-Type Hyperbranched Polymer to Polyamide-6 Fabrics", *Polymers Research Journal*, 7, 00, (2013).

45. Hua, Y., Zhang, F. Lin, H, and Chen. Y. *Application of Amino-terminated Hyperbranched Polymers in Reactive Dyeing on Silk Fabric*, Silk, (2008–09).

46. Zhang, F., Zhang, D. Chen, Y. and Lin, H. The antimicrobial activity of the cotton fabric grafted with an amino-terminated hyperbranched polymer. *Cellulose*, **16**, 281–288 (2009).

47. Burkinshaw, S. M., Mignanelli, Froehling, M. P. E., and Bide, M. J. The use of dendrimers to modify the dyeing behavior of reactive dyes on cotton. *Dyes and Pigments*, **47**, 259±267 (2000).

48. Gao, C., Xu, Y., Yan, D., and Chen, W. Water-Soluble Degradable Hyperbranched Polyesters: Novel Candidates for Drug Delivery?, *Biomacromolecules*, **4**, 704–712 (2003).

49. Wu, D., Liu, Y., Jiang, X., He, C., Goh, S. H., and Leong, K. W. Hyperbranched Poly (amino ester) with Different Terminal Amine Groups for DNA Delivery, *Biomacromolecules*, 7, 1879–1883 (2006).

50. Kima, T. H., Cooka, S. E., Arote, R. B., Cho, M. H., Nah, J. W., Choi, Y. J., and Cho, C. S., A Degradable Hyperbranched Poly (ester amine) Based on Poloxamer Diacrylate and Polyethylenimine as a Gene Carrier. *Macromol. Biosci.*, 7, 611–619 (2007)

51. Reul, R., Nguyen, J., and Kissel T., Amine-modified hyperbranched polyesters as non-toxic, biodegradable gene delivery systems. *Biomaterials*, **30**, 5815–5824 (2009).

52. Goswami, A., and Singh, A. K., Hyperbranched polyester having nitrogen core: synthesis and applications as metal ion extractant. *Reactive & Functional Polymers*, **61** 255–263 (2004).

53. Fang, J., Kita, H., and Okamoto, K., Hyperbranched Polyimides for Gas Separation Applications. 1. Synthesis and Characterization. *Macromolecules*, **33**, 4639–4646 (2000)

54. Seiler, M., Köhler, D., and Arlt, W., Hyperbranched polymers: new selective solvents for extractive distillation and solvent extraction. *Separation and Purification Technology*, **30** 179–197 (2003).

55. Itoh, T., Hirai, K., Tamura, M., Uno, T., Kubo, M., and Aihara, Y. Synthesis and characteristics of hyperbranched polymer with phosphonic acid groups for high-temperature fuel cells. *J Solid State Electrochem.*

56. Tang, W., Huang, Y., Meng, W., and Qing, F. L. Synthesis of fluorinated hyperbranched polymers capable as highly hydrophobic and oleophobic coating materials. *European Polymer Journal*, **46** 506–518 (2010).

57. Hong, X., Chen. Q., Zhang. Y., and Liu, G. Synthesis and Characterization of a Hyperbranched Polyol with Long Flexible Chains and Its Application in Cationic UV Curing. *Journal of Applied Polymer Science*, **77**, 1353–1356 (2000).

58. Shokrieh, M. M. and Rafiee, R. Prediction of Young's modulus of graphene sheets and carbon nanotubes using nanoscale continuum mechanics approach, *Materials and Design*, **31**,790–795 (2010).

59. Tserpes, K. I., and Papanikos, P. Finite element modeling of single-walled carbon nanotubes. *Composites: Part B*, **36**, 468–477 (2005).

60. Arani, A. G., Rahmani, R., and Arefmanesh, A. Elastic buckling analysis of single-walled carbon nanotube under combined loading by using the ANSYS software. *Physica E*, **40**, 2390–2395 (2008).

61. Ruoff, R. S., Qian, D., and Liu, W. K. Mechanical properties of carbon nanotubes: theoretical predictions and experimental measurements. *C. R. Physique*, **4**, 993–1008 (2003).

62. Guo, X., Leung, A. Y. T., He, X. Q., Jiang, H., and Huang, Y. Bending buckling of single-walled carbon nanotubes by atomic-scale finite element. *Composites: Part B*, **39**, 202–208 (2008).

63. Xiao. J. R., Gama, B. A., and Gillespie, J. W. An analytical molecular structural mechanics model for the mechanical properties of carbon nanotubes. *International Journal of Solids and Structures*, **42** 3075–3092 (2005).

64. Ansari, R., and Motevalli, B. The effects of geometrical parameters on force distributions and mechanics of carbon nanotubes: A critical study, *Commun. Nonlinear. Sci. Numer. Simulat.*, **14**, 4246–4263 (2009).

65. C. Li, T.W. Chou, Modeling of elastic buckling of carbon nanotubes by molecular structural mechanics approach. *Mechanics of Materials*, **36**, 1047–1055, (2004).

66. Natsuki, T., and Endo, M. Stress simulation of carbon nanotubes in tension and compression. *Carbon*, **42**, 2147–2151, (2004).

67. Ansari, R., Sadeghi, F., and Motevalli, B. A comprehensive study on the oscillation frequency of spherical fullerenes in carbon nanotubes under different system parameters. *Commun. Nonlinear. Sci. Numer. Simulat.*, **18**, 769–784, (2013).

68. Alisafaei, F., and Ansari, R. Mechanics of concentric carbon nanotubes: Interaction force and suction energy. *Computational Materials Science*, **50**, 1406–1413, (2011).

69. Natsuki, T., Ni, Q. Q., Endo, M. Vibrational analysis of fluid-filled carbon nanotubes using the wave propagation approach, *Appl. Phys. A*, **90**, 441–445 (2008).

70. Joshi, U. A., Sharma, S. C., and Harsha, S. P. Modeling and analysis of mechanical behavior of carbon nanotube reinforced composites. *Proc. IMechE, Part N: Journal of Nanoengineering and Nanosystems*, **225**, 23–31 (2011).

71. Shokrieh, M. M., and Rafiee, R. A review of the mechanical properties of isolated carbon nanotubes and carbon nanotube composites, *Mechanics of Composite Materials*, **46**, 155–172, (2010).

72. Sadus, R. J. *Molecular Simulation of Fluids: Algorithms and Object- Orientation*, Elsevier Amsterdam (1999).

73. Allen, M. P., and Tildesley, D. J. *Computer simulation in chemical physics*, Dordrecht Kluwer Academic Publishers (1993).

74. Haile, J. M. *Molecular dynamics simulation: elementary methods*, New York, Wiley (1997).

75. Liu, W. K., Karpov, E. G., and Park, H. S. *Nano Mechanics and Materials: Theory, Multiscale Methods and Applications*, John Wiley & Sons (2006)

76. Rapaport, D. C. The Art of Molecular Dynamics Simulation, Cambridge University Press New York (1995).

77. Aerts, J. Prediction of intrinsic viscosities of dendritic, hyperbranched and branched polymers. Computational and Theoretical Polymer Science, **8**, 49–54 (1998).

78. Lescanec, R. L. & Muthukumar, M. Configurational Characteristics and Scaling Behavior of Starburst Molecules: A Computational Study. *Macromolecules*, **23**, 2280–2288 (1990).

79. Le, Tu C., Todd, B. D., Daivis, P. J., and Uhlherr, A. Structural properties of hyperbranched polymers in the melt under shear via nonequilibrium molecular dynamics simulation. *The Journal of Chemical Physics* **130**, 074901 (2009).

80. Alexey, V. Lyulin, Adolf, David B., and Davies, Geoffrey R. Computer Simulations of Hyperbranched Polymers in Shear Flows. *Macromolecule*s, **34**, 3783–3789 (2001).

81. Konkolewicz, D., Thorn-Seshold, O., and Gray-Weale,A. Models for randomly hyperbranched polymers: Theory and simulation, *The Journal of Chemical Physics*, **129**, 054901 (2008).

82. De Gennes, P. G., and Hervet, H. J. Phys *Lett (Paris)*, **44**, L351 (1983).

83. Maiti, P, K, Cagın, T, Wang, G F, Goddard, W. A. *Macromolecules*, **37**, 6236–54 (2004).

84. Cagin, T., Wang, G., Martin, R., Breen, N., and Goddard III, W. A. Molecular modeling of dendrimers for nanoscale applications. *Nanotechnology*, **11**, 77–84 (2000).

85. Cagin, T., Wang, G., Martin, R., Zamanakos, G., Vaidehi, N., Mainz, D. T., and Goddard III, W. A., Multiscale modeling and simulation methods with applications to dendritic polymers. *Computational and Theoretical Polymer Science*, 00, 000±000 (2001).

86. Lue, L. Volumetric behavior of athermal dendritic polymers: Monte Carlo simulation. *Macromolecules*, **33**, 2266–2272 (2000).

87. Cam Le, T. *Computational Simulation of Hyperbranched Polymer Melts Under Shear*, Ph.D. thesis, Swinburne University of Technology (2010)

88. Zacharopoulos, N, and Economou. I. G. Morphology and organization of poly (propylene imine) dendrimers in the melt from molecular dynamics simulation, *Macromolecules*, (2002).

89. Lee, I., Athey, B. D., Wetzel, A.W., Meixner. W, and Baker, J. R. *Macromolecules*, **35**, 4510–20 (2002)

90. Hana, M., Chen, P., Yang, X. Molecular dynamics simulation of PAMAM dendrimer in aqueous solution. *Polymer*, **46**, 3481–3488 (2005).

91. Laferla, R. Conformations and dynamics of dendrimers and cascade macromolecules. *Journal of Chemical Physics*, **106**, 688–700 (1997).

92. Avila-Salas, F., Sandoval, C., Caballero, J., Guiñez-Molinos, S., Santos, L. S., Cachau,ll, R. E., and González-Nilo, F. D. Study of Interaction Energies between the PAMAM Dendrimer and Nonsteroidal Anti-Inflammatory Drug Using a Distributed Computational Strategy and Experimental Analysis by ESI-MS/MS, *J. Phys. Chem. B*, **116**, 2031–2039 (2012).

93. He, X., Liang, H., and Pan, C. Monte Carlo Simulation of Hyperbranched Copolymerizations in the Presence of a Multifunctional Initiator, *Macromol. Theory Simul.*, **10**, 196–203 (2001).

94. Neelov, I. M. and Adolf, D. B. Brownian dynamics simulation of hyperbranched polymers under elongational flow. *Journal of Physical Chemistry B*, **108**, 7627–7636 (2004).

95. Bosko, J. T. and Prakash, J. R. Effect of molecular topology on the transport properties of dendrimers in dilute solution at Theta temperature: A Brownian dynamics study. *Journal of Chemical Physics*, **128**, 034902 (2008).

96. Dalakoglou, G. K., Karatasos, K., Lyulin, S. V., and Lyulin, A. V. Shear-induced effects in hyperbranched-linear polyelectrolyte complexes. *Journal of Chemical Physics*, **129**, 034901 (2008).

97. Konkolewicz, D., Gilbert, G. R., and Gray-Weale, A. Randomly Hyperbranched Polymers. *Physical Review Letters*, **98**, 238301 (2007).

CHAPTER 3

EMULSION APPROACH TO MAGNETIC NANOCOMPOSITES

ALI GHARIEH

CONTENTS

ABSTRACT

Polymer nanocomposites are commonly defined as the combination of a polymer matrix and additives that have at least one dimension in the nanometer range. One of the most important fields which have gained an increasing interest in recent years is magnetic nanocomposites. In this review, the basics of magnetic properties of materials will be presented along with emulsion polymerization approach to magnetic latexes.

3.1 INTRODUCTION

Nanotechnology is one of the key technologies of the 21st Century, which is making great steps forward in the improvement of existing materials and the production of advanced and innovative materials based on both inorganic, organic materials as well as nanocomposites.

Polymer nanocomposites are commonly defined as the combination of a polymer matrix and additives that have at least one dimension in the nanometer range. The additives can be one-dimensional (examples include nanotubes and fibres), two-dimensional (which include layered minerals like clay), or three- dimensional (including spherical particles).

Over the past decade, polymer nanocomposites have attracted considerable interests in both academia and industry; this is because of the outstanding mechanical properties like elastic stiffness and strength which can be achieved with only a small amount of the nanoadditives. This is caused by the large surface area to volume ratio of nanoadditives when compared to the micro- and macro-additives. Other superior properties of polymer nanocomposites include barrier resistance, flame retardancy, scratch/wear resistance, as well as optical, magnetic and electrical properties.

The main difference between conventional fillers for polymers, such as clay, mica and talc, is the particle size and the aspect ratio of the filler. In the case of conventional fillers, the dispersed particles are relatively large aggregates of primary particles.

The benefits of nanocomposites compared to the polymers filled with traditional micrometer scale particles:

1. Increased modulus at very low filler concentration: 5 % nanofiller can typically provide the same increase in modulus as 40 % traditional mineral filler (such as talc) or 15 % glass fibres.
2. Lower density, because filler is used in much smaller quantities.
3. Increased barrier properties against gas transport through the nanocomposite. (Because of the high aspect ratio of filler.)
4. Reduced rate of moisture uptake in polymers such as polyamides, because of the barrier property.
5. Reduced flammability; the barrier function of the filler layers reduces the transport of oxygen and waste-gasses and filler char layer forms that blocks the burning polymer from the atmosphere.
6. Better surface and optical properties; due to the small particle size the surface of a nanocomposite is very smooth, and nanocomposites can be transparent because the particle size is below the wavelength of visible light.
7. An increased viscosity and often a melt yield stress.

3.2 OUTLINE OF THE CHAPTER

One of the most important fields which have gained an increasing interest in recent years is magnetic nanocomposites that will be discussed in this chapter of book. In the following sections, the basic of magnetic properties of materials will be presented briefly in section 3.3. In section 3.4, different conventional preparation methods for magnetic nanoparticles will be discussed. A brief discussion about the necessity of encapsulation of inorganic magnetic nanocomposites and their classification will be presented in section 3.5. Surface modification of inorganic magnetic nanoparticles will be discussed in section 3.6, and finally emulsion polymerization approach to magnetic latexes will be presented in section 3.7.

3.3 MAGNETIC PROPERTIES OF MATERIALS

Magnetic fields interact with all matter to some extent, so all materials are magnetic materials. We can classify materials depending on how they interact with magnetic fields as paramagnetic, diamagnetic, and ferromagnetic.

When a magnetic field, B, is applied to a material, the magnetic moments, m, will tend to align themselves either parallel to or anti-parallel to the applied field. Actually both effects occur in most materials but one is dominant. Materials dominated by the parallel alignment of m with external magnetic fields are called paramagnetic. Materials which display the anti-parallel alignment are diamagnetic. On a macroscopic scale, we can define the magnetization, M, which is the magnetic moment per unit volume. Another class of magnets remains magnetized even after the removal of the applied field. These are the ferromagnetic materials and are represented by the magnets we use in every day. It is suggested that these materials had domains, in which all the magnetic moments were aligned. Adjacent domains, however, were randomly oriented with respect to one another as indicated in Figure 1. In an applied magnetic field, the domain walls will move with domains whose magnetization is parallel with the applied field growing at the expense of other domains.

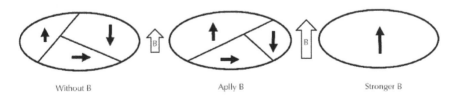

Without B   Aplly B   Stronger B

FIGURE 1   Effect of external magnetic field on the alignment of magnetic domains (m)

When the applied magnetic field is removed some of the domains lose their orientation, but the material does not return all the way to a random configuration. As a result it retains some magnetic properties; it has become a permanent magnet. The magnetic properties of these materials can be described by plotting a hysteresis loop for the magnetization, M, of the material as a function of the applied magnetic field,

B. Actually it is easier to measure the auxiliary magnetic field, H, so that is what is usually plotted. A typical plot is shown in Figure 2.

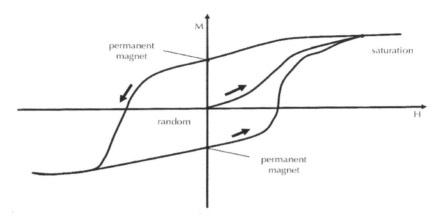

FIGURE 2   Hysteresis loop for the magnetization, M, of the material as a function of the applied magnetic field

The material starts out at the origin with grains randomly aligned producing no magnetization in no applied field. As the magnetic field is increased the domain walls move and the magnetization increases aligned with the field. Eventually all of the domains have magnetizations aligned with the field and the magnetization saturates. If the field is decreased the magnetization will slowly decrease as some of the domains lose their orientation with the field. When zero field is reached, the material still has some residual magnetization and has become a permanent magnet. If the direction of the field is now reversed, and increased in the opposite direction, the domain walls will again move to favor domains aligned with the new reversed field. This will change the magnetization down through zero and to a saturated state in the opposite direction. As this applied field is reduced to zero, some grains will again lose their orientation with the field, but a net magnetization will remain. The material is now a permanent magnet but in the opposite direction from before.

3.4   SYNTHESIS OF MAGNETIC NANOPARTICLE

Magnetic nanoparticles have been synthesized with a number of different compositions and phases, including iron oxides, such as Fe_3O_4 and γ-Fe_2O_3,[1-3] pure metals, such as Fe and Co,[4-7] spinel type ferromagnets, such as $NiFe_2O_4$, $MnFe_2O_4$[8-10], as well as alloys, such as $CoPt_3$[11,12] and $FePt$[13] .

In the last decades, much research has been devoted to the synthesis of magnetic nanoparticles. Several popular methods including micelle synthesis, co-precipitation, thermal decomposition and hydrothermal synthesis can all be directed at the synthesis of high-quality magnetic nanoparticles. In the following parts, the different methods which are common in preparation of magnetic nanoparticles will be discussed.

3.4.1 MICROEMULSION

A microemulsion is a thermodynamically stable isotropic dispersion of two immiscible liquids, where the microdomain of either or both liquids is stabilized by an interfacial film of surfactant molecules. In water-in-oil microemulsions, the aqueous phase is dispersed as microdroplets (typically 1–50 nm in diameter) surrounded by a monolayer of surfactant molecules in the continuous hydrocarbon phase. The size of the reverse micelle is determined by the molar ratio of water to surfactant[14].

By mixing two identical water-in-oil microemulsions containing the desired reactants, the microdroplets will continuously collide, coalesce, and break again, and finally a precipitate forms in the micelles. By the addition of solvent, such as acetone or ethanol, to the microemulsions, the precipitate can be extracted by filtering or centrifuging the mixture. In this sense, a microemulsion can be used as a nanoreactor for the formation of nanoparticles[14].

Using the microemulsion technique, metallic cobalt[15], cobalt/platinum alloys[16], and gold-coated cobalt/platinum nanoparticles[17] have been synthesized. MFe_2O_4 (M: Mn, Co, Ni, Cu, Mg, or Cd, etc.) are among the most important magnetic materials and have been widely used for electronic applications[18,19].

Spinel ferrites can be synthesized in microemulsions and inverse micelles. For example, $MnFe_2O_4$ nanoparticles with controllable sizes from about 4–15 nm are synthesized through the formation of water-in-toluene inverse micelles with sodium dodecylbenzenesulfonate (NaDBS) as surfactant[20]. This synthesis starts with a clear aqueous solution consisting of $Mn(NO_3)_2$ and $Fe(NO_3)_3$. A NaDBS aqueous solution is added to the metal salt solution, subsequent addition of a large volume of toluene forms reverse micelles. The volume ratio of water and toluene determines the size of the resulting $MnFe_2O_4$ nanoparticles.

Magnetic nanoparticles of cobalt ferrite ($CoFe_2O_4$) have been synthesized using water-in-oil microemulsions consisting of water, cetyltrimethyl ammonium bromide (surfactant), n-butanol (cosurfactant), and n-octane (oil). Precursor hydroxides were precipitated in the aqueous cores of water-in-oil microemulsions and these were then separated and calcined to give the magnetic oxide[21].

Although many types of magnetic nanoparticles have been synthesized in a controlled manner using the microemulsion method, the particle size and shapes usually vary over a relative wide range. Large amounts of solvent are necessary to synthesize appreciable amounts of material. It is thus not a very efficient process and also rather difficult to scale-up.

3.4.2 CO-PRECIPITATION

Co-precipitation is a facile and convenient way to synthesize iron oxides (either Fe_3O_4 or γ-Fe_2O_3) from aqueous Fe^{2+}/Fe^{3+} salt solutions by the addition of a base under inert atmosphere at room temperature or at elevated temperature [22,23]. The size, shape, and composition of the magnetic nanoparticles very much depends on the type of salts used (e.g. chlorides, sulfates, nitrates), the Fe^{2+}/Fe^{3+} ratio, the reaction temperature, the pH value and ionic strength of the media. With this synthesis, once the synthetic conditions are fixed, the quality of the magnetite nanoparticles is fully reproducible.

Magnetite nanoparticles are not very stable under ambient conditions, and are easily oxidized to maghemite (Fe_2O_3, γ-Fe_2O_3) or dissolved in an acidic medium. Since maghemite is a ferrimagnet, oxidation is the lesser problem. Therefore, magnetite particles can be subjected to deliberate oxidation to convert them into maghemite. This transformation is achieved by dispersing them in acidic medium, then addition of iron(III) nitrate. The maghemite particles obtained are then chemically stable in alkaline and acidic medium.

However, even if the magnetite particles are converted into maghemite after their initial formation, the experimental challenge in the synthesis of Fe_3O_4 by co-precipitation lies in control of the particle size and thus achieving a narrow particle size distribution. Since the blocking temperature depends on particle size, a wide particle size distribution will result in a wide range of blocking temperatures and therefore non ideal magnetic behavior for many applications [24].

3.4.3   THERMAL DECOMPOSITION

Monodisperse magnetic nanocrystals can essentially be synthesized through the thermal decomposition of organometallic compounds in high-boiling organic solvents containing stabilizing surfactants.

In literature the various type of organometallic precursors like metal acetylacetonates[25], metal cupferronates [26] were used. In addition several types of surface active agents like, or carbonyls, Fatty acids, oleic acid, hexadecylamine and, $C_6H_5N(NO)O$-) were used as surfactants [27].

In principle, the proportions of the starting reagents including organometallic compounds, surfactant, and solvent are the key parameters for the control of the size and morphology of magnetic nanoparticles. The reaction temperature, reaction time, as well as aging period may also be essential for the exact control of size and morphology [28,29].

If the metal in the precursor is zerovalent, such as in carbonyls, thermal decomposition initially leads to formation of the metal, but two-step procedures can be used to produce oxide nanoparticles as well [30]. In a related work, the synthesis of highly crystalline and monodisperse γ-Fe_2O_3 nanocrystallites is reported. High-temperature (300°C) aging of iron-oleic acid metal complex, which was prepared by the thermal decomposition of iron pentacarbonyl in the presence of oleic acid at 100°C, was found to generate monodisperse iron nanoparticles [31].

The decomposition of precursors with cationic metal centers leads directly to the oxides, in the following of the latter mentioned report [31], the resulting iron nanoparticles were transformed to monodisperse γ-Fe_2O_3 nanocrystallites by controlled oxidation by using trimethylamine oxide as a mild oxidant. Particle size can be varied from 4 to 16 nm by controlling the experimental parameters.

3.4.4   HYDROTHERMAL SYNTHESIS

Under hydrothermal conditions a broad range of nanostructured materials can be shaped. Hydrothermal process uses water as solvent and takes advantage from pressure and temperature to increase the solubility of the precursors and speed up the reaction time. The hydrothermal synthesis can be carried out above or below the supercritical point of water [32-34]. Supercritical water can provide an excellent reaction environment for hydrothermal crystallization of metal oxide particles. Because of the drastic change of properties of water around the critical point, density, dielectric constant, and ionic product, the phase behavior for the supercritical water−light gas (O_2, H_2, etc.) system and reaction equilibrium/rate can be varied to synthesize new materials or define particle morphologies. A suitable reactor or a sealed reaction vessel is needed to control the temperature and the pressure during the reaction. This synthesis rout is able to produce mono disperse particles with good crystallinity [35,36].

3.5 MAGNETIC NANOCOMPOSITE

3.5.1 INTRODUCTION

The basic hint in a composite is to incorporate several component materials and their properties in a single material. In magnetic nanocomposites, organic-inorganic synergies add new properties that cannot be achieved in just organic or inorganic materials by themselves. Bare magnetic nanoparticles are sensitive to oxidation in air therefore it is necessary to develop efficient strategies to avoid any stability issues. This can be achieved by the production of a polymeric shell, which will not only protect the inorganic component, but will also provide the nanoparticles with selective functionalities needed for further applications. The magnetic part of composite has nanometer dimensions, therefore its magnetic properties can differ qualitatively from the bulk, and they vary with the particle size. The polymeric part, brings interesting mechanical, optical, and electrical properties besides the high processability. In addition the magnetic properties can be strongly affected by the polymer interphase, and the capacity of polymers for structuring can be used to control the interparticle magnetic interactions

To date, the major field of interest remains the biomedical field [37,38], in which the magnetic nanoparticles have been successfully used as solid support for the purification [39], extraction [40], and concentration of biomolecules [41], as contrast agents in magnetic resonance imaging [42], as mediators in hyperthermia43, and as carriers for guided drug delivery [44].

3.5.2 CLASSIFICATION OF MAGNETIC POLYMER NANOCOMPOSITES

Magnetic polymer nanocomposites can be in the form in various types of shapes like: particulates, fibers, films and multilayers, or tridimensional solids. In each of these forms, they can be used in a large number of applications consequently with their properties. A scheme of this classification is shown in Figure 3. In the following parts we will discuss about the particulates magnetic nanocomposites. In order to produce these kinds of magnetic nanocomposites, emulsion method is one of the most outstanding techniques which will be reviewed in following sections.

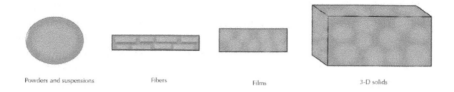

Powders and suspensions   Fibers   Films   3-D solids

FIGURE 3   Different shapes of magnetic polymeric nanocomposites

3.6 SURFACE MODIFICATION

Because of the hydrophobic nature of monomers and hydrophilic nature of magnetic nanoparticles, prior to carrying out emulsion process, the surface of the magnetic nanoparticles must be converted into a more hydrophobic one by using surface modifiers or coupling agents. The most important goals of surface modification are: introduction of different reactive groups useful in the subsequent polymerization reaction and enhancement of the surface hydrophobicity of the inorganic particles.

For the encapsulation of hydrophilic magnetic nanoparticles, the surfaces have to be converted into more hydrophobic by functionalization with a surfactant having a low hydrophilic/lipophilic balance (HLB) value, with coupling agents or surface modifiers. Apart from the enhancement of the surface hydrophobicity of the inorganic particles, in some cases different reactive groups are introduced for polymerization with the hydrophobic monomer [45].

In the case of magnetic nanoparticles several surface modifiers were used [46-51], between them fatty acids, such as oleic acid (OA), oleoyl sarcosine acid (OSA) or stearic acid are extensively used [52-55].

3.7 EMULSION POLYMERIZATION

Heterogeneous polymerization, especially emulsion polymerization, provides an effective way of synthesizing nanoparticles with various architectures and forms. In the case of an emulsion polymerization, homogeneous and micellar nucleations are the main mechanisms for particle formation. In the presence of magnetic nanoparticles dispersed in the aqueous phase, particles surface can be an additional site for nucleation. Thus, the control of the morphology of the composite nanoparticles can become complicated due to the competition among these nucleation mechanisms.

The preparation of magnetic latex particles using emulsion polymerization in the presence of a freshly prepared ferrofluid was first reported in the late 1970s and at the beginning of the 1980s but was not investigated in detail [56,57]. Since then, a great number of studies have been published in the literature, and magnetic nanoparticles are one of the most documented types of inorganic particle being used to form composite colloids. An overview of the various methods reported in the literature is given in the following sections.

3.7.1 SURFACTANT BILAYER (ADMICELLAR POLYMERIZATION)

One of the basic requirements for efficient encapsulation of inorganic nanoparticles is to enhance the interfacial affinity between the nanoparticles and the monomer. One frequently encountered strategy for achieving this is to create hydrophobic loci inside a bilayer of surfactant(s) (figure 4). Surfactant is the one coating the nanoparticles after their synthesis and allowing dispersion of the nanoparticles in nonpolar solvents. Once the excess of the primary surfactant is removed, the nanoparticles are coated with a secondary surfactant (not necessary to be the same as the first surfactant) to form a self-organized bilayer of the two surfactants on the surface of the nanoparticles, thus allowing their dispersion in water [58-60]. The hydrophobic interlayer thus formed between the two surfactants can solubilize the monomer and finally promote the polymerization close to/at the vicinity of the surface of the nanoparticles, according to the so-called admicellar polymerization mechanism.

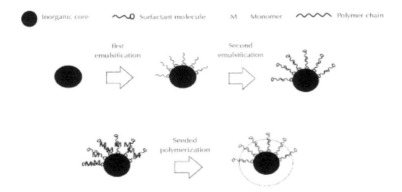

FIGURE 4 Magnetic nanoparticle encapsulation through admicellar polymerization.

Meguro et al. were among the first to explore this method for the encapsulation of non-magnetic iron oxide (α-Fe_2O_3) and titanium dioxide through emulsion polymerization of styrene adsolubilized into adsorbed bilayers of sodium dodecyl sulfate (SDS) [61]. Using the same concept, magnetic PS and PMMA particles were obtained by Yanase et al. [62] using a commercial ferrofluid with magnetite particles covered by sodium oleate and sodium dodecylbenzenesulfonate (SDBS). The process yielded PS particles with up to 20 wt% of encapsulated Fe_3O_4.

In other attempt, aqueous magnetic fluids were synthesized by a sequential process involving the chemical coprecipitation of Fe(II) and Fe(III) salts with ammonium hydroxide (NH_4OH) followed by resuspension of the ultrafine particles in water using fatty acid[58]. This procedure produced Fe_3O_4 nanoparticles stabilized against agglomeration by bilayers of n-alkanoic acids with 9–13 carbons encapsulating the metal particles. In other report, bilayer surfactant coating on magnetite (Fe_3O_4) nanoparticles has been obtained using the self-assembly method[59].

In other work [60], Magnetic fluids consisting of magnetite nanoparticles and a surrounding bilayer of a primary and a secondary fatty acid surfactant were prepared using either 10-undecenoic acid or undecanoic acid for one or both of the surrounding layers. The olefin units were included within the structure as sites for polymerizing the shell components and increasing the stability of the magnetic fluid. The magnetic fluids were exposed to various levels of γ irradiation and observed that the position of the unsaturated fatty acid in the bilayer affected their extent of polymerization, as evidenced by IR and NMR spectroscopies. When 10-undecenoic acid was used solely for the primary layer, γ irradiation resulted in only 50% conversion of the olefinic groups; however, their complete conversion occurred when 10-undecenoic acid was used as both the primary and secondary surfactants of the bilayer. Light scattering measurements showed that this latter magnetic fluid displayed a significant improvement in stability after irradiation on dilution with water. Gel permeation chromatography provided a measure of the aggregation number for the polymerized surfactant shell within these various fluids and showed that the polymerization process was affected by the irradiation dose as well as by the inclusion of the unsaturated surfactant as either the primary or secondary layer or as both layers of the bilayer coating.

Magnetic latex particles of PMMA (in the range of 100 ± 50nm) were obtained by soapless seeded emulsion polymerization performed in the presence of 10-nm Fe_3O_4 nanoparticles coated with a bilayer of lauric acid [63]. This work shed light on the importance of keeping a good balance between the amount of iron oxide nanoparticles (and hence the surfactant bilayer) and the initial amount of MMA: too high an amount of monomer (higher than the bilayer could accommodate, thus leading to destruction of the bilayer) led to the expected seeded emulsion polymerization but also to a crop of particle generated by self-nucleation (including either homogeneous or micellar nucleation).

In an approach very similar to Wang's work, γ-Fe_2O_3 modified by myristic acid and soluble in octane was dispersed in SDS solution [64]. Following polymerization of styrene, divinylbenzene (DVB) and NaSS provided composite particles, but iron oxide nanoparticles were limited to the surface of the polymer particles. These particles nevertheless easily aligned in the presence of an external magnetic field and could find potential applications in proton-exchange membranes.

Preparation of magnetic nano-carriers possessed uniform core/shell/shell nanostructure composed of 40 nm magnetite particles/poly(styrene-co-glycidyl methacrylate (GMA))/polyGMA, which was constructed by admicellar polymerization was reported by Honda et al.[65]

The admicellar polymerization concept was also applied to the synthesis of thermosensitive magnetic latex particles based on N-isopropylacrylamide (NIPAM) [66,67]. In this case, however, the polymerization could be better defined as seeded precipitation polymerization owing to the water solubility of this monomer.

3.7.2   OTHER SURFACE COATINGS OF IRON OXIDE NANOPARTICLES BY OTHER METHODS

The surfactant bilayer strategy is obviously not the only method that has been developed to favor polymerization at the surface of iron oxide nanoparticles. Thus, other (macro) molecules have been employed to this aim. Various polymers such as poly acrylic acid68, poly(ethylene glycol) [69] or dextran derivatives [70] have been used as steric stabilizers to form aqueous dispersions of iron oxide nanoparticles for use in emulsion polymerization. In the case of poly(ethylene glycol) (PEG), fluorescent and magnetic polysaccharide-based particles were prepared in three steps [71]. First, commercial magnetite powder and europium phthalate complex (fluorescent) were blended and dispersed in a PEG solution to obtain fluorescent magnetite colloid particles (FMCPs). Copolymerization of styrene and maleic anhydride in the presence of FMCPs seeds led to magnetite europium phthalate/poly(styrene-co-maleic anhydride) core–shell composite microspheres. Finally, heparin was conjugated with the surface anhydrides to form FMCPs/SMA heparin glycoconjugate core–shell composite particles.

In another work, commercial Fe_3O_4 was modified with PEG for the synthesis of azidocarbonyl-functionalized magnetic particles via a two-step procedure [45]. First, magnetic poly(styrene-co-AAm-co-AA) particles were obtained through emulsion polymerization performed in water/ethanol mixture in the presence of PEG-modified Fe_3O_4 and a small amount of SDS. Azidocarbonyl groups were then converted into amido groups and successfully used for covalent protein immobilization.

In a report that was published by Wu et al. [72], In order to overcome the low conversion and complex post-treatment, four different polymerization procedures were adopted to prepare the magnetic polymer latexes. The results clearly show that the strategy using magnetic emulsion template-dosage is the most effective and feasible. Based on the optimized procedure, various factors including the type of initiators such as oil soluble initiator, water soluble initiator, redox initiator system, crosslinking agent, functional monomers etc. were systematically studied. Magnetic polymer latex with high monomer conversion of 83% and high magnet content of 31.8% was successfully obtained. Besides, core–shell structured magnetic polymer latex with good film forming property was also prepared, which is promising for potential applications such as magnetic coatings and modification of cementitious materials with controlled polymer location.

3.8 CONCLUSION

In order to reach magnetic polymeric nanocomposites, emulsion polymerization technique could be one of the best choices. Inorganic magnetic nanoparticles must be produced by appropriate methods, in some applications uniformity of shape and size is crucial for final uses and synthesis method must be chosen by enough knowledge about the selected techniques. Inorganic nanoparticles which are produced have super paramagnetic characteristic. To take more advantage of this valuable property, it is necessary to create a polymeric shell on the surface of these nanoparticles. In order to reach the highest yield in encapsulation step, the surface modification of inorganic magnetic nanoparticles is not preventable. In the final step, polymeric shell is formed on the surface of inorganic nanoparticles via emulsion polymerization, therefore magnetic polymeric nanocomposites are produced.

KEYWORDS

- **Emulsion polymerization**
- **Magnetic nanocomposites**
- **Polymer nanocomposites**

REFERENCES

1. Casula, M. F., Corrias, A., Arosio, P., Lascialfari, A., Sen, T., Floris, P., and Bruce, I. J. *Journal of Colloid and Interface Science*, **357**(1), 50–55 (2011).
2. Prithviraj Swamy, P. M., Basavaraja, S., Lagashetty, A., Srinivas Rao, N. V., Nijagunappa, R., and Venkataraman, A. *Bull Mater Sci*, **34**(7), 1325–1330 (2011).
3. Sun, S., Anders, S., Hamann, H. F., Thiele, J. U., Baglin, J. E. E., Thomson, T., Fullerton, E. E., Murray, C. B., and Terris, B. D. *Journal of the American Chemical Society*, **124**(12), 2884–2885 (2002).
4. Chang, Y. C. and Chen, D. H. *Journal of Colloid and Interface Science*, **283**(2), 446–451 (2005).
5. Boyen, H. G., Kästle, G., Zürn, K., Herzog, T., Weigl, F., Ziemann, P., Mayer, O., Jerome, C., Möller, M., Spatz, J. P., Garnier, M. G., and Oelhafen, P. *Advanced Functional Materials*, **13**(5), 359–364 (2003).
6. Crangle, J. and Goodman, G. M. Proceedings of the Royal Society of London. *A. Mathematical and Physical Sciences*, **321**(1547), 477–491 (1971).
7. Fert, A. and Piraux, L. *Journal of Magnetism and Magnetic Materials*, **200**(1–3), 338–358 (1999).
8. Liu, C., Zou, B., Rondinone, A. J., and Zhang, Z. J. *The Journal of Physical Chemistry B*, **104**(6), 1141–1145 (2000).
9. Vestal, C. R. and Zhang, Z. J. *Journal of the American Chemical Society*, **124**(48), 14312–14313 (2002).
10. Shi, W., Zhang, X., He, S., and Huang, Y. *Chemical Communications*, **47**(38), 10785–10787 (2011).
11. Park, J. I. and Cheon, J. *Journal of the American Chemical Society*, **123**(24), 5743–5746 (2001).
12. Shevchenko, E. V., Talapin, D. V., Schnablegger, H., Kornowski, A., Festin, Ö., Svedlindh, P., Haase, M., and Weller, H. *Journal of the American Chemical Society*, **125**(30), 9090–9101 (2003).
13. Song, T., Zhang, Y., Zhou, T., Lim, C. T., Ramakrishna, S., and Liu, B. *Chemical Physics Letters*, **415**(4–6), 317–322 (2005).
14. López Pérez, J. A., López Quintela, M. A., Mira, J., Rivas, J., and Charles, S. W. *The Journal of Physical Chemistry B*, **101**(41), 8045–8047 (1997).
15. Chen, J. P., Lee, K. M., Sorensen, C. M., Klabunde, K. J., and Hadjipanayis, G. C. *Journal of Applied Physics*, **75**(10), 5876–5878 (1994).
16. Kumbhar, A., Spinu, L., Agnoli, F., Wang, K. Y., Zhou, W., and O'Connor, C. J. *Magnetics, IEEE Transactions on*, **37**(4), 2216–2218 (2001).
17. O'Connor, C. J., Sims, J. A., Kumbhar, A., Kolesnichenko, V. L., Zhou, W. L., and Wiemann, J. A. *Journal of Magnetism and Magnetic Materials*, 226–230, Part 2, (0), 1915–1917 (2001).
18. Zaki, T., Saed, D., Aman, D., Younis, S. A., and Moustafa, Y. M. *J Sol-Gel Sci Technol*, **65**(2), 269–276 (2013).

19. Zhang, Z., Rondinone, A. J., Ma, J. X., Shen, J., and Dai, S. *Advanced Materials*, **17**(11), 1415–1419 (2005).
20. Liu, C., Zou, B., Rondinone, A. J., and Zhang, Z. J. *The Journal of Physical Chemistry B*, **104**(6), 1141–1145 (2000).
21. Pillai, V. and Shah, D. O. *Journal of Magnetism and Magnetic Materials*, **163**(1–2), 243–248 (1996).
22. Wei, H. and Wang, E. *Analytical Chemistry*, **80**(6), 2250–2254 (2008).
23. Jeong, J. R., Lee, S. J., Kim, J. D., and Shin, S. C. *physica status solidi (b)*, **241**(7), 1593–1596 (2004).
24. Qu, Y., Yang, H., Yang, N., Fan, Y., Zhu, H., and Zou, G. *Materials Letters*, **60**(29–30), 3548–3552 (2006).
25. Seo, W. S., Jo, H. H., Lee, K., Kim, B., Oh, S. J., and Park, J. T. *Angewandte Chemie International Edition*, **43**(9), 1115–1117 (2004).
26. Ghosh, M., Biswas, K., Sundaresan, A., and Rao, C. N. R. *Journal of Materials Chemistry*, **16**(1), 106–111 (2006).
27. Banerjee, R., Katsenovich, Y., Lagos, L., McIintosh, M., Zhang, X., and Li, C. Z. *Current Medicinal Chemistry*, **17**(27), 3120–3141 (2010).
28. Demortiere, A., Panissod, P., Pichon, B. P., Pourroy, G., Guillon, D., Donnio, B., and Begin-Colin, S. *Nanoscale*, **3**(1), 225–232 (2011).
29. Ocaña, M., Morales, M. P., and Serna, C. J. *Journal of Colloid and Interface Science*, **171**(1), 85–91 (1995).
30. Lu, A. H., Salabas, E. L., and Schüth, F. *Angewandte Chemie International Edition*, **46**(8), 1222–1244 (2007).
31. Hyeon, T., Lee, S. S., Park, J., Chung, Y., and Na, H. B. *Journal of the American Chemical Society*, **123**(51), 12798–12801 (2001).
32. Adschiri, T., Hakuta, Y., and Arai, K. *Industrial & Engineering Chemistry Research*, **39**(12), 4901–4907 (2000).
33. Cabanas, A. and Poliakoff, M. *Journal of Materials Chemistry*, **11**(5), 1408–1416 (2001).
34. Adschiri, T. *Chemistry Letters*, **36**(10), 1188–1193 (2007).
35. Ma, J., Lian, J., Duan, X., Liu, X., and Zheng, W. *The Journal of Physical Chemistry C*, **114**(24), 10671–10676 (2010).
36. Li, X., Si, Z., Lei, Y., Tang, J., Wang, S., Su, S., Song, S., Zhao, L., and Zhang, H. *CrystEngComm*, **12**(7), 2060–2063 (2010).
37. Gao, J., Gu, H., and Xu, B. *Accounts of Chemical Research*, **42**(8), 1097–1107 (2009).
38. Hao, R., Xing, R., Xu, Z., Hou, Y., Gao, S., and Sun, S. *Advanced Materials*, **22**(25), 2729–2742 (2010).
39. Soelberg, S. D., Stevens, R. C., Limaye, A. P., and Furlong, C. E. *Analytical Chemistry*, **81**(6), 2357–2363 (2009).
40. Huang, C. and Hu, B. *Spectrochimica Acta Part B: Atomic Spectroscopy*, **63**(3), 437–444 (2008).
41. Gu, H., Ho, P. L., Tsang, K. W. T., Wang, L., and Xu, B. *Journal of the American Chemical Society*, **125**(51), 15702–15703 (2003).
42. Cunningham, C. H., Arai, T., Yang, P. C., McConnell, M. V., Pauly, J. M., and Conolly, S. M. *Magnetic Resonance in Medicine*, **53**(5), 999–1005 (2005).
43. Johannsen, M., Gneveckow, U., Eckelt, L., Feussner, A., WaldÖFner, N., Scholz, R., Deger, S., Wust, P., Loening, S. A., and Jordan, A. *International Journal of Hyperthermia*, **21**(7), 637–647 (2005).
44. Veiseh, O., Gunn, J. W., and Zhang, M. *Advanced Drug Delivery Reviews*, **62**(3), 284–304 (2010).

45. Pich, A., Bhattacharya, S., Ghosh, A., and Adler, H. J. P. *Polymer*, **46**(13), 4596–4603 (2005).
46. Bruce, I. J. and Sen, T. *Langmuir*, **21**(15), 7029–7035 (2005).
47. Liu, X., Guan, Y., Ma, Z., and Liu, H. *Langmuir*, **20**(23), 10278–10282 (2004).
48. Zhang, Y., Kohler, N., and Zhang, M. *Biomaterials*, **23**(7), 1553–1561 (2002).
49. Bourlinos, A. B., Bakandritsos, A., Georgakilas, V., and Petridis, D. *Chemistry of Materials*, **14**(8), 3226–3228 (2002).
50. Mikhaylova, M., Kim, D. K., Bobrysheva, N., Osmolowsky, M., Semenov, V., Tsakalakos, T., and Muhammed, M. *Langmuir*, **20**(6), 2472–2477 (2004).
51. Rong, M. Z., Zhang, M. Q., Wang, H. B., and Zeng, H. M. *Applied Surface Science*, **200**(1–4), 76–93 (2002).
52. Zhang, L., He, R., and Gu, H. C. *Applied Surface Science*, **253**(5), 2611–2617 (2006).
53. Zhao, Y. X., Zhuang, L., Shen, H., Zhang, W., and Shao, Z. J. *Journal of Magnetism and Magnetic Materials*, **321**(5), 377–381 (2009).
54. Qiu, G., Wang, Q., Wang, C., Lau, W., and Guo, Y. *Polymer International*, **55**(3), 265–272 (2006).
55. Hong, R. Y., Pan, T. T., and Li, H. Z. *Journal of Magnetism and Magnetic Materials*, **303**(1), 60–68 (2006).
56. Rembaum, A., Yen, S. P. S., and Molday, R. S. *Journal of Macromolecular Science: Part A - Chemistry*, **13**(5), 603–632 (1979).
57. Bourgeat-Lami, E. and Lansalot, M., Organic/Inorganic Composite Latexes: The Marriage of Emulsion Polymerization and Inorganic Chemistry. In *Hybrid Latex Particles*, A. M. Herk, and K. Landfester (Eds.) Springer Berlin Heidelberg, **233**, 53–123 (2010).
58. Shen, L., Laibinis, P. E., and Hatton, T. A. *Langmuir*, **15**(2), 447–453 (1998).
59. Fu, L., Dravid, V. P., and Johnson, D. L. *Applied Surface Science*, **181**(1–2), 173–178 (2001).
60. Shen, L., Stachowiak, A., Hatton, T. A., and Laibinis, P. E. *Langmuir*, **16**(25), 9907–9911 (2000).
61. Nayyar, S. P., Sabatini, D. A., and Harwell, J. H. *Environmental Science & Technology*, **28**(11), 1874–1881 (1994).
62. Yanase, N., Noguchi, H., Asakura, H., and Suzuta, T. *Journal of Applied Polymer Science*, **50**(5), 765–776 (1993).
63. Wang, P. C., Chiu, W. Y., Lee, C. F., and Young, T. H. *Journal of Polymer Science Part A: Polymer Chemistry*, **42**(22), 5695–5705 (2004).
64. Brijmohan, S. B. and Shaw, M. T. *Journal of Membrane Science*, **303**(1–2), 64–71 (2007).
65. Nishio, K., Masaike, Y., Ikeda, M., Narimatsu, H., Gokon, N., Tsubouchi, S., Hatakeyama, M., Sakamoto, S., Hanyu, N., Sandhu, A., Kawaguchi, H., Abe, M., and Handa, H. *Colloids and Surfaces B: Biointerfaces*, **64**(2), 162–169 (2008).
66. Shamim, N., Hong, L., Hidajat, K., and Uddin, M. S. *Colloids and Surfaces B: Biointerfaces*, **55**(1), 51–58 (2007).
67. Cao, Z., Ziener, U., and Landfester, K. *Macromolecules*, **43**(15), 6353–6360 (2010).
68. Lee, C. F., Chou, Y. H., and Chiu, W. Y. *Journal of Polymer Science Part A: Polymer Chemistry*, **45**(14), 3062–3072 (2007).
69. Qu, F., Guan, Y., Ma, Z., and Zhang, Q. *Polymer International*, **58**(8), 888–892 (2009).
70. Molday, R. S. and Mackenzie, D. *Journal of Immunological Methods*, **52**(3), 353–367 (1982).
71. Qiu, G. M., Xu, Y. Y., and Zhu, B. K., Qiu, G. L. *Biomacromolecules*, **6**(2), 1041–1047 (2005).
72. Wu, C. C., Kong, X. M., and Yang, H. L. *Journal of Colloid and Interface Science*, **361**(1), 49–58 (2011).

CHAPTER 4

THE FRACTAL KINETICS OF POLYMERIZATION CATALYZED BY NANOFILLERS

G. V. KOZLOV and G. E. ZAIKOV

CONTENTS

ABSTRACT

The fractal analysis of polymerization kinetics in nanofiller presence was performed. The influence of catalyst structural features on chemical reaction course was shown. The notions of strange (anomalous) diffusion conception was applied for polymerization reactions description.

4.1 INTRODUCTION

By Sergeev's definition the nanochemistry is a science field connected with obtaining and studing of physical-chemical properties of particles having sizes of nanometer scale. Let's note that according to this definition polymers synthesis is automatically a nanochemistry part as far as according to the Melikhov's classification polymeric macromolecules (more precisely macromolecular coils) belong to nanoparticles and polymeric sols and gels – to nanosystems. Catalysis on nanoparticles is one of the most important sections of nanochemistry.

The majority of catalytic systems are nanosystems. At heterogeneous catalysis the active substance is tried to deposit on carrier in nanoparticles form in order to increase their specific surface. At homogeneous catalysis active substance molecules have often in them selves nanometer sizes. The most favorable conditions for homogeneous catalysis are created when reagent molecules are adsorbed rapidly by nanoparticles and are desorbed slowly but have high surface mobility and, consequently, high reaction rate on the surface and at the reaction molecules of such structure are formed at which desorption rate is increased sharply. If these conditions are realized in nanosystem with larger probability than in macrosystem, then nanocatalyst has the raising activity that was observed for many systems. In the connection such questions arise as adsoption and desorption rate, surface mobility of molecules and characteristics frequency of reagents interaction acts depend on the size, molecular relief and composition of nanoparticles and the carrier.

The present chapter purpose it the application of fractal analysis for description of polymerization kinetics in nanofiller presence.

4.2 RESULTS AND DISCUSSION

The analysis of polymerization processes in nanofiller presence does not differ principally from the one for transesterification model reaction [1]. In the present chapter some important aspects of such polymerization will be studied, mainly on the example of solid-phase imidization.

The authors [2] studied the kinetics of poly (amic acid) (PAA) solid-state imidization in nanofiller (Na^+-montmorillonite) presence and found an essential reaction acceleration both at imidization temperature T_i raising in the range 423-523 K and Na^+-montmorillonite contens W_c increase within range 0–7 phr. The possible chemical mechanism of Na^+-montmorillonite action as a catalyst was offered, assuming larger contact surface area and respectively larger number of reaction active sites, that promotes dehydration and imide ring closure reaction [2]. This model has hypothetical and qualitative character. However, it became obvious lately, that on chemical reactions kinetics large influence can be exercised by purely physical factors such as re-actionary medium connectivity [3] or macromolecular coil structure [4], which in all polymer's states (solution, melt, solid phase) is fractal [5, 6]. It is also well known [7], that the fractal objects description is correct only within the framework of fractal geometry and the usage in such case of Euclidean geometry gives approximation more or less corresponding to reality. Proceed from the said above the authors [8, 9] exercised the solid-state imidization reaction description both in nanofiller presence and in its absence within the framework of structural (fractal) models.

In the general and the simplest form solid-state imidization reaction can be represented by the equation [3]:

$$A + A \rightarrow \text{inert product} , \qquad (1)$$

Where A is a reagent (in considered case PAA).

Then for such reaction description the following relationship was used [3]:

$$\rho_A \sim t^{-d_s/2} , \qquad (2)$$

where ρ_A is the concentration of nonreacted reagent A, which further will be accepted equal to $(1-Q)$ (Q is a conversion degree), t is a reaction duration, d_s is a spectral dimension.

In Figure 1 the dependences $\rho_A = (1-Q)$ on t in log-log coordinates, corresponding to the relationship (2), for imidization reaction without filler at the four indicates above imidization temperatures $T_i = 423, 473, 503$, and 523 K are shown.

As follows from the data of Figure 1, all the four adduced plots are linear, that allows to determine the value of spectral dimension d_s. The estimations have shown, that the imidization temperature T_i raising within the range 423–523 K results to d_s increase from 0.42 up to 1.68, i.e. to essential growth of reactionary system connectivity degree. In Fig. 2 the similar dependences for various Na$^+$-montmorillonite contents W_c at fixed $T_i = 473$ K are shown. As one can see, nanofiller introduction exercises much weaker influence on d_s value than the imidization temperature raising [9].

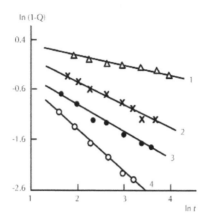

FIGURE 1   The dependences of $\rho_A = (1-Q)$ on t in log-log coordinates, corresponding to the rela-tionship (2), for PAA solid-state imidization without filler at temperatures: 423 (1), 473 (2), 503 (3) and 523 (4) [8].

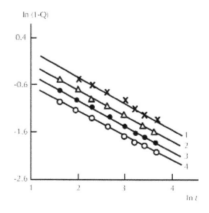

FIGURE 2   The dependences of $\rho_A = (1-Q)$ on t in log-log coordinates, corresponding to the relationship (2), for PAA solid-state imidization at temperature 473 K and Na^+-montmorillonite contents W_c: 0 (1), 2 (2), 5 (3) and 7 (4) phr [8].

d_s increase at T_i raising for the same reactionary system, shown in Figure 1, assumes, that in the considered case d_s should be considered as an effective spectral dimension

d_s', depending on reactionary medium heterogeneity degree [10]. The medium heterogeneity degree can be characterized by heterogeneity exponent h, which changes within the range $0<h<1$ and turns into zero only for homogeneous samples [10]. The values h and d_s' are connected with one another by the equation [10]:

$$d_s' = 2(1-h).$$ (3)

In Figure. 3 the dependence $h(T_i)$ is shown, from which fast decrease h or reactionary medium homogeneity raising follows at T_i increase. At $T_i{\approx}540$ K the exponent $h=0$, i.e. the reactionary medium becomes homogeneous. The authors [2] have shown, that for the studied polyimides the melting temperature T_m is equal about to 800 K. Proceeding from the known law of two-thirds [11]:

$$\frac{T_g}{T_m} = \frac{2}{3},$$ (4)

The glass transition temperature T_g of polyimide can be estimated as equal to ~ 533 K. In other words, as it was expected [12, 13], reactionary medium in solid-state imidization case became homogeneous (Euclidean) at glass transition.

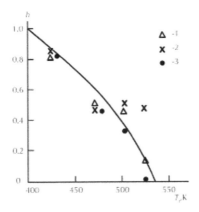

FIGURE 3 The dependence of reactionary medium heterogeneity exponent h on imidization temperature T_i for PAA solid-state imidization at Na^+-montmorillonite contents W_c: 0 (1), 2 (2) and 5 (3) phr [8].

The shape of the curve $h(T_i)$, shown in Fig. 3, i.e. tendency for $h{\to}0$ at temperature raising, assumes, that the fractal-like effects, namely, d_s' variation, are connected with energetic disorder [10].

Let us consider physical principles of reactionary system connectivity degree change, characterized by effective spectral dimension d_s', at imidization temperature

T_i and Na$^+$-montmorillonite contents W_c change. As well as earlier, the value of macromolecular coil fractal dimension D_f can be estimated with the help of formula [4]:

$$t^{(D_f-1)/2} = \frac{c_1}{k_1(1-Q)},$$ (5)

where c_1 is constant, determined according to the boundary conditions and accepted equal to $8' 10^{-4}$ s^{-1} for studied reactions, k_1 is reaction rate constant.

In Figure. 4 the dependence of d'_s on D_f is shown, from which it follows, that d'_s increases at D_f reduction and at $D_f \rightarrow 1.50$ (transparent macromolecular coil [5]) the value d'_s has a fast tendency to its limiting magnitude $d'_s = 2.0$ [10]. Such form of dependence $d'_s(D_f)$ allows to make two conclusions. Firstly, the definite interconnection of d'_s and D_f characterized by the curve of Figure 4 exists. Secondly, the value d'_s cannot be considered as spectral dimension of proper macromolecular coil, since in this case theory assumes d'_s decrease at D_f reduction [14]:

$$d'_s = \frac{2(2D_f - d)}{d+2},$$ (6)

where d is dimension of Euclidean space, in which fractal is considered (it is obvious, that in our case $d = 3$).

Proceeding from the said above, let us assume, that in the considered case d'_s is a reactionary medium connectivity indicator, in some way connected with macromolecular coil structure characterized by dimension D_f. Let us consider one of the possible theoretical schemes of such interrelation.

It is well known [15], that in chemical reactions large effect has steric factor p ($p£1$) showing that not all collisions of reagents occur with proper for reaction products formation orientation of reacting molecules.

The value p is defined by dimension D_f and can be calculated according to the equation [4]:

$$p = \frac{1.6}{10600^{(D_f-1)/2}}.$$ (7)

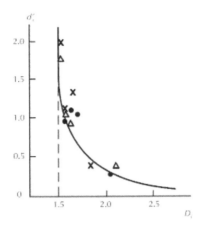

FIGURE 4   The dependence of reactionary medium effective spectral dimension d_s' on macromolecular coil fractal dimension D_f in imidization process. The notation is the same, as in Figure. 3 [8].

As it follows from this equation, the value p is increased at D_f reduction. Macromolecular coil sites number N, capable to take part in a chemical reaction (active sites), is determined like that [16]:

$$N \sim t^{d_s/2} . \qquad (8)$$

Let us pay attention, that $N \sim \rho_A^{-1}$ in the relationship (2). But in case of a chemical reaction not all active sites of macromolecular coil can react in virtue of the condition $p<1$, but only their part N_p, proportional to p. For preliminary estimations it can be assumed [9]:

$$N_p = 100p . \qquad (9)$$

Believing that for imidization reaction without filler at $T_i = 423$ K $D_f = 2.12$, d_s' =0.42, i.e. experimentally determined values, one can calculate the value p according to the equation (7), the value N_p according to the equation (9) and to determine constant coefficient in the relationship (8) at t = const = 15 min., which is equal to 2. Further, using this coefficient in the equation (8), one can calculate the values d_s' , which further will be designated as $(d_s')_{th}$, for imidization reaction with variable T_i and W_c. In Figure 5 the comparison of values d_s', determined from the slope of plots $\rho_A(t)$ in log-log coordinates (see Figs. 1 and 2) and $(d_s')_{th}$, calculated according to the equations (7)-(9), is shown. As one can see, between these parameters a good correspondence was obtained. This means, that reactionary space connectivity, character-

ized by dimension d_s', depends on macromolecular coil structure dimension D_f and this dimension is specific namely for the chemical reactions owing to steric factor p introduction [9].

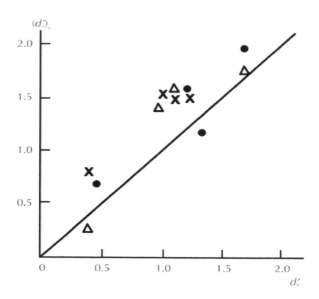

FIGURE 5 The comparison of the calculated according to the relationship (2) d_s' and the equations (7)-(9) $(d_s')_{th}$ effective spectral dimension values of reactionary medium in imidization process. The notation is the same, as in Figure 5.3 [8].

From the point of view of process chemistry this can be treated as follows: the smaller D_f is the more open macromolecular coil structure, is the easier dehydration process (water removal) and imidic ring closure proceed [2].

Therefore, the data considered above demonstrated that the main parameter, controlling solid-state imidization rate, is the reactionary system connectivity degree characterized by its effective spectral dimension. In its turn, this dimension is a function of macromolecular coil structure that is polymeric reaction specific feature. Imidization temperature raising defines reactionary medium heterogeneity reduction and corresponding increase of its connectivity degree [8,9].

The quantitative analysis of the imidization kinetics temperature dependence was given within the framework of one more conception, namely, a chemical reactions kinetics fractal model [4,17,18]. The authors [17] have assumed that the cause of imidization reaction acceleration at its temperature T_i growth is polyimide macromolecular coil structure change, which is the consequence of its molecular characteristic [17] ratio C_∞ was chosen, which is a polymer chain statistical flexibility indicator [19]. As earlier, the macromolecular coil structure is characterized by its fractal (Hausdorff) dimension D_f. The interrelation between C_∞ and D_f is given by the equation [20]:

$$C_\infty = \frac{D_f}{3(2.28 - D_f)} + \frac{4}{3}.$$ (10)

The temperature dependence C_∞ (and, consequently, D_f) can be calculated, using the equation offered in paper [13]:

$$C_\infty = \left(1 - \sqrt{1 - \frac{9}{8e}\frac{T_m}{T_i}}\right) / const,$$ (11)

where T_m is the melting temperature, for the studied nanocomposites equal to ~ 800 K [2].

For the constant in the equation (11) determination the following method were used. The general equation of chemical reactions fractal kinetics is the formula [4]:

$$Q = c_0 \eta_0 t^{(3-D_f)/2},$$ (12)

where c_0 is reagent initial concentration, η_0 is initial viscosity of reactionary medium.

By constructing the dependence $Q(t)$ in log-log coordinates at $T_i = 523$ K the value $D_f = 1.59$ was determined, then from the equation (10) corresponding to it the value C_∞ and according to the equation (11) – the value of estimating constant. Further for the remaining imidization temperatures the values C_∞ were calculated (the general variation 3.51–4.51) and corresponding to them the values D_f (the general variation 1.59–1.81). Then according to the equation (12) the kinetic curves $Q(t)$ were calculated. The value of reactionary medium initial viscosity η_0 in this case was accepted constant and equal to 1 (with taking into account of the fact that the reaction occurs in solid phase) and the value c_0 was determined by the selection method at the condition of the best correspondence of theoretical and experimental curves $Q(t)$. Simulation of kinetic curves of solid-state imidization carried out by the considered method is shown by points in Figure 6. As one can see that a good correspondence of theory and experiment is obtained.

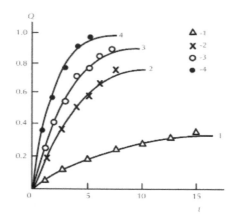

FIGURE 6 The kinetic curves $Q(t)$ of solid-state imidization at temperatures: 423 (1, 5), 473 (2, 6), 503 (3, 7) and 523 K (4, 8). 1-4 – experimental data; 5-8 – the calculation according to the equation (12) [17].

The selection of reagents initial concentration (or reaction active centers) shows it increase within the range 6.5–35 of relative units at imidization temperature T_i growth within the range 423–523 K. This c_0 increase can also be explained within the framework of chemical reactions kinetics fractal conception according to which [4]:

$$M \sim c_0 t_{gen},   (13)$$

where M is molecular weight of reaction product, t_{gen} is reaction duration.

Assuming M = const, we obtain theoretical value $c_0 (c_0^{th})$ according to the relationship (13) [17]:

$$c_0^{th} \sim t_{gen}^{-1}.   (14)$$

Accepting as t_{gen} imidization reaction part duration, on which the first order reaction laws are fulfilled and on which imidization mainly was ended, the authors [17] have estimated the values c_0^{th}, which are compared with the values c_0, obtained by a selection method, in Figure 7. As it follows from the data of this Figure, between the values c_0^{th} and c_0 the linear correlation is observed, passing through coordinates origin. Such correspondence assumes, that the value c_0 (imidization active centers number) increases at T_i raising at the expense of diffusive processes intensification [4].

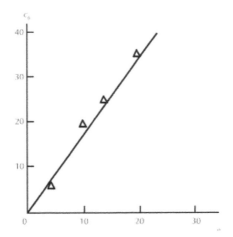

FIGURE 7   The relation of by selection obtained c_0 and calculated according to the relationship (14) c_0^{th} number of imidization reaction active sites [17].

Hence, the stated above results have shown, that the fractal conception of chemical reactions kinetic describes quantitatively kinetics of solid-phase imidization process at different temperatures. This description is given only within the framework of reaction physical aspects and does not affect its chemical aspects. The effective initial concentration of reagents c_0 at imidization temperature growth is due to physical cause also – by reagents diffusion intensification in solid-phase state.

It is necessary to indicate that the nanofiller introduction in reactionary mixture results to two-phase system formation, where an important (or decisive) role will be played by interfacial interactions [21]. Particularly, the interaction of PAA-Na$^+$-montmorillonite should result to the structure change of polyimide (PI) forming macromolecular coil [22] and similar effect gives the imidization temperature T_i raising [23]. Therefore the authors [24] fulfilled structural analysis of processes, occurring in solid-phase imidization reaction course, according to the aspect indicated above.

Let us consider the interfacial interactions problem of PI forming macromolecular coil and Na$^+$-montmorillonite on nanofiller surface. As Pfeifer shows [22], a macromolecular coil on hard surface changes its configuration (structure), which can be characterized by its fractal dimension D_f.

This change is described with the help of the following equation [22]:

$$\frac{d_{surf} D_f^{sol}}{D_f} = d_{surf}^0, \qquad (15)$$

where d_{surf} and d_{surf}^0 are fractal dimensions of nanofiller surface in nanocomposite and in initial state, respectively, D_f^{sol} and D_f are fractal dimensions of PI macromolecular coil in solution (the blending of PAA and Na$^+$-montmorillonite was carried out in N,N-dimethylacetamide solution [2]) and in solid-phase state on nanofiller surface, respectively.

Let us consider the estimation of the parameters including in the equation (15). As it was shown in paper [25], a polymer chain, possessing by finite rigidity and consisting of statistical segments of finite length, was not capable to reproduce growing surface roughness at d_{surf}^0 increase and at $d_{surf}^0 > 2.5$ the value d_{surf} is determined as follows [26]:

$$d_{surf} = 5 - d_{surf}^0 . \tag{16}$$

For Na$^+$-montmorillonite the value d_{surf}^0 is determined experimentally and equal to 2.78 [27]. The value D_f^{sol} can be accepted in the first approximation equal to macromolecular coil dimension in a good solvent ($D_f^{sol} = 1.667$ [5]). Then the estimation according to the equation (15) gives D_f=1.33. It is obvious, that this dimension of the macromolecular coil, stretched on Na$^+$-montmorillonite surface will be designated further as D_f^0.

The calculation of real values of macromolecular coil fractal dimension D_f for the first order reaction, which is solid-state imidization [2], can be fulfilled with the help of the equation (5). In other words the calculation according to this equation shows that for the studied imidization reactions the condition $D_f^0 < D_f$ is fulfilled. Such relation allows assuming that only part of PI macromolecular coils interacts with Na$^+$-montmorillonite surface. This is confirmed by the data of Figure 8, where the difference $\Delta D_f = D_f - D_f^0$ is plotted on the graph as a function of nanofiller contents W_c for four imidization temperatures. As it follows from this Figure plots, the ΔD_f value decreases at W_c growth or $D_f = D_f^0$ and these plots extrapolation shows, that at $W_c \approx 17.5$ mass. % $D_f = D_f^0$ or ΔD_f=0. Let's note, that the indicated value W_c is true only for exfoliated (nonaggregated) nanofiller.

The ΔD_f decrease at W_c growth assumes the interacting phase fraction φ_{int} increase in the imidization process. The value φ_{int} can be determined according to the mixtures law from the equation [24]:

$$D_f = \phi_{int} D_f^0 + (1 - \phi_{int}) D_f' , \tag{17}$$

where D_f' is the macromolecular coil fractal dimension in nanofiller absence.

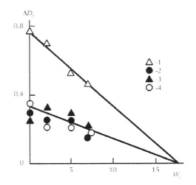

FIGURE 8 The dependences of the fractal dimension difference $\Delta D_f = D_f - D_f^0$ on nanofiller contents W_c for nanocomposites polyimide/Na$^+$-montmorillonite at imidization temperatures: 423 (1), 473 (2), 503 (3) and 523 K (4) [24].

In Figure 9 the dependence $\varphi_{int}(W_c)$ for $T_i = 423$ K is adduced. As one can see, this correlation is linear, passes through coordinates origin and is described analytically by the following empirical equation [24]:

$$\phi_{int} = 0.0575W_c, \qquad (18)$$

where W_c is given in mass. %.

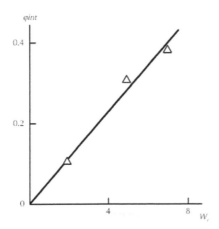

FIGURE 9 The dependence of interacting phase relative fraction φ_{int} on nanofiller contents W_c for nanocomposites polyimide/Na$^+$-montmorillonite at imidization temperature 423 K [24].

It is obvious, that at $W_c = 17.5$ mass%, obtained by plots of Figure. 8 extrapolation, the value $\varphi_{int} = 1.0$, i.e. in an imidization reaction the entire reactionary system PAA-Na$^+$-montmorillonite is influenced.

In Figure 10 the dependence of reaction rate constant k_1 on interacting phase relative fraction φ_{int}, is adduced which turns out approximately linear and shows k_1 growth at φ_{int} increase. This allows to assume the direct dependence of solid-phase imidization rate on interfacial interactions level in the reactionary system [24].

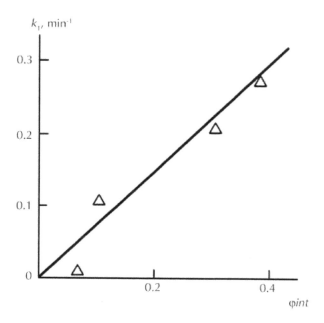

FIGURE 10   The dependence of the first order reaction rate constant k_1 on relative fraction of interacting phase φ_{int} for nanocomposites polyimide/Na$^+$-montmorillonite at imidization temperature 423 K [24].

The reduction of imidization process activation energy E_{act} at W_c increase was found out – from 66 up to 51 kj/mole within the range $W_c = 0$–7 mass% [2]. Earlier the authors [28] offered the following dependence of E_{act} on D_f in case of polyarylate thermooxidative degradation:

$$E_{act} = 16.6D_f^2 - 2.8D_f .$$ (19)

In Table 1 the comparison of experimental E_{act} and calculated according to the equation (19) E_{act}^{th} activation energy values of solid-state imidization is adduced. As one can see, a good correspondence between the indicated values of activation energy is obtained (the average discrepancy of E_{act} and E_{act}^{th} makes less than 5%). This means, that association energy (imidization reaction) and dissociation one (thermooxidative degradation) are approximately equal, that was to expected.

TABLE 1 The comparison of experimental E_{act} and calculated according to the equation (19) E_{act}^h values of solid-state imidization process activation energy [24]

W_c, mass. %	E_{act}, kj/mole	E_{act}^h, kj/mole
0	66.0	68.6
2	57.5	63.4
5	54.0	52.8
7	51.5	49.3

Hence, the results obtained above have demonstrated again that the cause of imidization process acceleration at nanofiller contents growth is macromolecular coil structure change owing to its interfacial interactions with Na^+-montmorillonite surface. The interacting phase relative fraction increases at nanofiller contents raising and at its content about 17.5 mass% this phase ocuppies the entire reactionary system. The imidization process activation energy reduction at nanofiller contents increase is also due to structural factors, namely, to a macromolecular coil fractal dimension decrease.

The authors [2] have found out that the kinetic curves $Q(t)$ have typical shape for polymerization reactions with auto deceleration showing imidization rate reduction as time is passing (see Fig. 6). Such curves $Q(t)$ are specific for the reaction course in heterogeneous medium and are described by a simple relationship [10]:

$$\frac{dQ}{dt} \sim t^{-h}, \tag{20}$$

where h is heterogeneity exponent ($0<h<1$), turning into zero for homogeneous (Euclidean) mediums; incidentally the behavious is classical: dQ/dt=const.

The mentioned relationship supposes strong effect of this heterogeneity degree characterized by exponent h on reaction rate. Therefore the authors [28,29] undertake an attempt of clarification of the reactionary medium heterogeneity physical significance in case of PAA solid-phase imidization and the factors defining the medium heterogeneity exponent value.

The solid-phase imidization reaction in the most simple and general form can be represented by the relationship (1), which can be described by the scaling relationship (2) for diffusion-limited reactions. In Figure 1 the dependences $\rho_A(t)$ in log-log coordinates, corresponding to the relationship (2), for solid-phase imidization reaction without filler at the four imidization temperatures T_i are shown. As it was noted above, the obtained dependences are linear and according to their slope the value of spectral dimension d_s characterized reactionary medium connectivity in the relationship (2) can be obtained. T_i increase within the range 423–523 K results to substantial growth of d_s: from 0.42 up to 1.68. Let us note that such d_s increase occurs without reactionary mixture composition change. This means, that the energetic restrictions result to the appearance of fractal space, in which instead of the value d_s an effective spectral dimension d_s' must be used, reflecting the existence of the restrictions mentioned above and connected with d_s by the equation [10]:

$$d_s' = \beta_j d_s, \tag{21}$$

where β_j is the parameter, characterizing distribution of reagents "jumps" (displacements) times.

In Figure 3 the dependence $h(T_i)$ is shown, from which fast decrease h or reduction of reactionary medium heterogeneity at T_i raising follows. At $T_i \approx 540$ K exponent $h = 0$, i.e. reactive medium becomes homogeneous. Since for PI $T_g \approx 533$ K, then, as it was expected [12], that the reactionary medium in case of solid-phase imidization became homogeneous (Euclidean) at glass transition. The shape of the curve $h(T_i)$, adduced in Figure 3, i.e. h goes to zero at temperature raising, assumes, that the fractal-like effects, namely, d_s' variation, are connected with energetic disorder [10]. In such case the energetic state of polymer structure can be characterized by an excess energy localization regions dimension D_f^e [30]. The value D_f^e can be estimated according to the following equation [31]:

$$D_f^e = \frac{4\pi T_i}{\ln\left(1/f_g\right) T_g}, \tag{22}$$

where f_g is a relative fluctuational free volume.

In Figure 11 the dependence $h(D_f^e)$ is adduced, from which the expected result: follows polymer structure energetic excitation degree raising, due to thermal energy "pumping" at T_i increase, results to h reduction. At $D_f^e \approx 6.3$ the reactionary medium becomes homogeneous ($h = 0$).

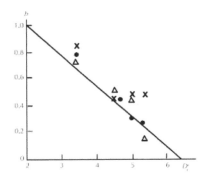

FIGURE 11 The dependence of heterogeneity exponent h of reactionary medium on excess energy localization regions dimension D_f^e for PAA solid-phase imidization at Na$^+$-montmorillonite contents W_c: 0 (1), 2 (2) and 5 (3) mass% [29].

Therefore, the data of Figure. 3 and 11 give the answer to the question, at what conditions $h=0$, i.e. when the reactionary medium becomes homogeneous. Nevertheless, the physics of this process remains vague. The glass transition gives singularities neither in f_g behavior nor in D_f^e behavior. Therefore for the explanation of heterogeneous↔homogeneous medium transition let's use representations of the conception of fractal (local) free volume D_f^e. According to this conception free volume microvoid is necessary to simulate not by three-dimensional sphere, as it was accepted in classical polymer physics [32], but by D_f^e-dimensional sphere with the volume v_h^{fr}. The value v_h can be estimated as follows [31]:

$$v_h^{1/3} = \left(\frac{T_m - T_i}{T_m} \right)^{-\nu},$$

(23)

where percolation index ν was accepted equal to 0.85 [33].

Further from geometrical considerations in the assumption of three-dimensional microvoid of free volume its radius r_h can be estimated and then v_h^{fr} can be calculated according to the equation [29]:

$$v_h^{fr} = \frac{\pi^{D_f^e/2} r_n^{D_f^e}}{\left(D_f^e / 2 \right)!},$$

(24)

where r_n is radius of free volume microvoid.

In Figure. 12 the dependence $h(f_g^{fr})$ is adduced where value f_g^{fr} was calculated according to the equations (22), (23) and

$$f_g^{fr} = f_g \frac{v_h^{fr}}{v_h} , \tag{25}$$

where relative fraction of fluctuational free volume f_g can be accepted equal to 0.060 for solid-phase polymers [12].

As it follows from the data of this Figure, value $h=0$ or reactionary medium homogeneity at $f_g^{fr}=0.34$ is achieved. Let's remind that the mentioned value f_g^{fr} corresponds to percolation threshold for overlapping spheres [34]. In other words, at f_g^{fr} =0.34 fluctuational free volume microvoids, simulated by D_f – dimensional sphere, form continuous percolation network or continuous diffusion channels [35]. Therefore, between heterogeneous and homogeneous reactionary medium, at any rate, in case of solid-phase imidization, qualitative difference exists. For heterogeneous reactionary medium dehydration product (water molecule), which is in a free volume microvoid, is forced to expect the opening of overlapping it neighboring microvoid, after that it makes "jump" from the first to the second and further the process repeats. For homogeneous reactionary medium such process of "expectation" is not required by virtue of the existence of through percolation channels of free volume. Let's note, that the mentioned processes of "jumps" are realized on local level. The indicated effect is the cause of diffusive processes intensification in solid-phase imidization course, which was mentioned above.

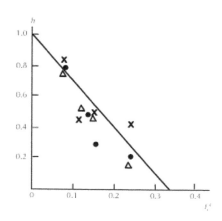

FIGURE 12   The dependence of heterogeneity exponent h of reactionary medium on relative fractal free volume f_g^{fr} for PAA solid-phase imidization. The notation is the same, that in Figure 11 [29].

And lastly, in Figure 13 the dependence of coefficient β_j in the equation (5.16) on f_g^{fr} is adduced. Again the value β_j reaches its limiting magnitude $\beta_j=1$ (i.e. $d_s'=d_s$) at $f_g^{fr}=0.34$. The relationship between β_j and f_g^{fr} is given by the simple empirical equation [29]:

$$\beta_j = 2.94 f_g^{fr}. \tag{26}$$

The plot of Figure 13 demonstrates that the energetic restriction, defining transition from d_s to d_s', is the necessity of "jumps" of reaction product or reagents between free volume microvoids. It is clear, that T_i raising decreases "jump" expectation time and the formation of through percolation channels of free volume microvoids cancels these restrictions.

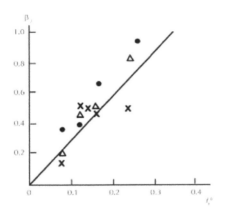

FIGURE 13   The dependence of the coefficient β_j in the equation (21) on relative fractal free volume f_g^{fr} for PAA solid-phase imidization. The notation is the same that in Figure. 11 [29].

Hence, the results considered above demonstrated that the notion of reactionary medium heterogeneity in case of solid-phase imidization was connected with free volume representations, that was expected for diffusion-limited solid-phase reactions. If free volume microvoids are not connected with one another, then medium is heterogeneous, and in case of formation of overlapping percolation network of such microvoids it is homogeneous. To obtain such definition is possible only within the framework of the fractal free volume conception.

The temperature imidization T_i raising within the range 423–523 K and the nanofiller contents W_c increase within the range 0–7 mass% results to essential imidization kinetics change expressed by two aspects: by an essential increase of reaction rate (reaction rate constant of the first order k_1 increases almost on two orders) and by raising of conversion (imidization) limiting degree Q_{lim} almost: from 0.25 for imidization reaction without nanofiller at $T_i = 423$ K up to 1.0 at Na^+-montmorillonite content 7 mass. % and $T_i = 523$ K [2]. Let us also remind, that all kinetic curves $Q(t)$ for the indicated imidization reactions have typical shape of curves with autodeceleration (see Fig. 6), characteristic for fractal reactions, i.e. either fractal objects reactions, or reactions in fractal spaces [36]. In other words, the indicated imidization reaction aspects in sufficient degree have general character. If for the first effect (k_1 increase) the authors [2] offered probable chemical treatment considering nanofiller as a catalyst, then the second effect (Q_{lim} raising) did not obtain any explanation, although its theoretical and practical significance is obvious. Therefore the authors [37] offered structural treatment of limiting conversion degree in solid-phase imidization process based on the general principles of fractal analysis [36].

A macromolecular coil in various polymer's states (solution, melt, solid phase) presents itself the fractal object characterized by fractal (Hausdorff) dimension D_f. Specific feature of fractal objects is the distribution of their mass in space: The density ρ of such object changes at its radius R variation as follows [5,6,38]:

$$\rho = \rho_{dens}\left(\frac{R}{a}\right)^{D_f - d},\qquad(27)$$

where ρ_{dens} is the density of material, which consists of fractal object in dense packing assumption, a is a lower linear scale of object fractal behavior, d is the dimension of Euclidean space, in which fractal is considered (it is obvious, that in our case $d = 3$).

From the equation (27) ρ decrease at D_f reduction follows, since it's always $D_f < d$, that, naturally, simplifies reagents access in macromolecular coil internal regions and results to the fuller chemical transformations, i.e. to conversion degree Q_{lim} increase. Besides, it is known [38] that at macromolecular coil formation by irreversible aggregation mechanisms in its central part densely-packed region is formed where the proceeding of reaction is impossible. Proceeding from that, it is possible to confirm, that for a chemical reaction only a part of macromolecular coil is accessible, which is the larger, the smaller its fractal dimension D_f is. In the transparent coil case ($D_f £ 1.5$ [5]) both low- and high-molecular substances can pass freely through it and this assumes, that in such case the value $Q_{lim} = 1.0$. At $D_f = 2.5$ chemical reaction ceases and gelation process begins [39,40]. This means, that at reaching $D_f = 2.5$ $Q_{lim} = 0$. The indicated estimations allow to write the fractional exponent v_f for chemical reactions similarly to the definition [41]:

$$v_f = D_f - \left(D_f^{gel} - 1\right) = D_f - 1.5,\qquad(28)$$

where D_f^{gel} is the value D_f at gelation, equal to 2.5.

Let us remind, that according to [42] the value v_f characterizes the system states fraction unchanging in its evolution process. In case of chemical reactions generally and imidization process particularly this assumes, that the value v_f characterizes the macromolecular coil part inaccessible for chemical transformations. Then the coil part β_{ac} accessible for such transformations is determined as follows [43]:

$$\beta_{ac} = 1 - v_f = 2.5 - D_f. \tag{29}$$

Proceeding from the said above, it's possible to define the limiting conversion degree Q_{lim} by the following identity [43]:

$$Q_{lim} = \beta_{ac}. \tag{30}$$

Therefore, the estimation Q_{lim} brings to the question of fractal dimension D_f determination. As it was noted above, at present two methods of indicated dimension determination exist. The first method consists of using chemical reactions fractal kinetics general relationship, i.e. the equation (12). Besides, the calculation of dimension D_f allows to have the equation (5).

Plotting the dependences $Q(t)$ in log-log coordinates allows to determine the value D_f according to the slope of these dependences in their linearity case according to the equation (12). In Fig. 14 the mentioned dependences for process of PAA imidization without nanofiller are shown. As one can see, these dependences are linear, that allows to make estimation D_f by the indicated method.

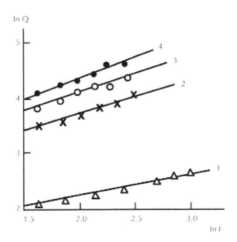

FIGURE 14 The dependences of imidization degree Q on reaction duration t in log-log coordinates for PAA imidization process at temperatures: 423 (1), 473 (2), 503 (3) and 523 K (4) [37].

In Figure 15 the comparison of the dimensions D_f calculated by two methods (D_{f1} and D_{f2}, respectively) is shown. As one can see, both methods give close values D_f and therefore further their average magnitude will be used, i.e. $D_f = (D_{f1} + D_{f2})/2$.

Further the parameter β_j can be estimated according to the equation (29) and compared with the limiting conversion degree Q_{lim}, obtained experimentally [2]. Such comparison for PAA imidization process without nanofiller and in the presence of 2 mass. % of Na⁺-montmorillonite at four temperatures of imidization indicated above is shown in Figure 16. Good enough correspondence of theory and experiment (their average discrepancy is equal to ~ 12 %) was obtained, that confirms the offered treatment correctness [37].

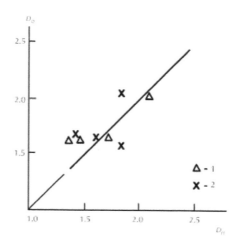

FIGURE 15   The comparison of macromolecular coil fractal dimension D_{f1} and D_{f2} calculated according to the relationships (12) and (5), respectively, for PAA imidization process without filler (1) and in the presence of 2 mass% of Na⁺-montmorillonite (2) [37].

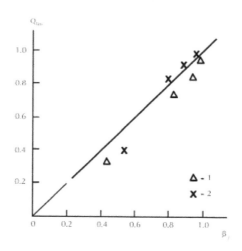

FIGURE 16   The dependence of limiting imidization degree Q_{lim} on parameter β_j value for PAA imidization process without nanofiller (1) and in the presence of 2 mass% of Na⁺-montmorillonite (2) [37].

Hence, the data stated above assume, that limiting conversion (in the given case – imidization) degree is defined by purely structural parameter – macromolecular coil fraction, subjected to the evolution (transformation) in chemical reaction course. This fraction can be correctly estimated within the framework of fractal analysis. For this purpose two methods of macromolecular coil fractal dimension calculation have been offered, which give coordinated results.

It is known, that the majority of catalytic systems are nanosystems [44]. At the heterogeneous catalysis active substance is being tried to deposite on the carries in nanoparticles form in order to increase their specific surface. At homogeneous catalysis active substance molecules have often by them selves nanometer sizes. It is known too [2] that the operating properties of heterogeneous catalyst systems depend on their geometry and structure of surface, which can influence strongly on catalytic properties, particularly, on catalysis selectivity. It was shown [27] that the montmorillonite surface is a fractal object. Proceeding from this, the authors [47] studied the montmorillonite fractal surface effect on its catalytic properties in isomerization reaction.

Two types of montmorillonite were used – Na- montmorillonite (SW) and Ca-montmorillonite (ST) [47]. The indicated types of layered silicate were applied as a catalyst at isomerization of 1-butene (B) by obtaining of *cis*-2-butene (C) and trans-2-butene (T).

Meakin [42] considered the simplest catalysis scheme, which was used later for estimation of catalyst selectivity S_c. It demonstrates general features, inherent to all catalysis models, and is expressed by a simple reaction scheme, which by analogy with the reactions (2.16)-(2.21) can be written as follows [46]:

$$A + P \rightarrow A_a, \tag{31}$$

$$A \xrightarrow[(cat)]{k_1} B, \tag{32}$$

$$A + A_a \xrightarrow[(cat)]{k_2} C, \tag{33}$$

where the reaction (31) presents itself molecule A adsorption on catalyst surface P. The reaction (32) presents itself unimolecular process, transformating adsorbed molecule A(A_a) in new molecule B, which is assumed as fastly leaving the catalyst surface. In real systems this can be an isomerization reaction (as in the considered case) or a reaction of secondary products decay. The reaction (33) presents itself molecule A addition to catalyst surface in the site already, occupied by the adsorbed molecule A(A_a) with subsequent reaction of molecule C formation, which is also assumed by fastly leaving this surface in order to make the model maximally simple. The selectivity S_c is determined as molecules C number, divided on molecules B number [45].

Within the framework of this model with using computer simulation the following expression for S_c was obtained [45]:

$$S_c = \frac{k_f \Sigma_i P_i^2 / \left(k_1 + 2P_i k_f\right)}{k_1 \Sigma_i P_i / \left(k_1 + 2P_i k_f\right)}, \tag{34}$$

where k_f is molecules A receiving rate on catalyst surface, P_i is contact probability for i-th site of the indicated surface.

The equation (34) is simplified essentially for two limiting cases. In the large k_1 limit it accepts the form [45]:

$$S_c = \frac{k_f}{k_1} \Sigma_i P_i^2, \tag{35}$$

and in the small k_1 limit let us obtain the equation

$$S = \frac{k_f}{Nk_1},$$

in which general number of surface sites N can be estimated according to the following general fractal relationship [48]:

$$N \sim L^{d_{surf}}, \tag{36}$$

where L is characteristic size of nanofiller particle, accepted equal to 100 nm [2], d_{surf} is catalyst surface fractal dimension.

In its turn, d_{surf} value can be estimated with the help of the relationship [49]:

$$S_u = 410 \left(\frac{L}{2}\right)^{d_{surf}-d}, \tag{37}$$

where S_u is a specific surface of nanofiller particles, d is dimension of Euclidean space, in which fractal is considered (it is obvious, that in our case $d=3$). The value S_u is given in m²/g and L – in nm.

In Figure 17 the dependence $S_c(N)$ is adduced for the two types of the studied catalyst – montmorillonite. In case of 1-butene isomerization reaction the equations (31)-(33) can be written as follows [46]:

$$B + P \rightarrow B_a, \tag{38}$$

$$B_a \xrightarrow[\text{(cat)}]{k_1} C, \tag{39}$$

$$B + B_a \xrightarrow[\text{(cat)}]{k_2} T, \tag{40}$$

and then the value S_c is defined as the ratio T/C.

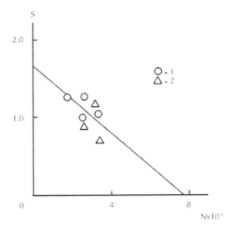

FIGURE 17 The dependence of the catalyst selectivity S_c on its surface sites general number N for Na- montmorillonite (1) and Ca- montmorillonite (2) [46].

From Figure 17 data S_c linear decrease at N growth follows. According to the equation (31) this assumes isomerization reaction course at small k_1 and at fulfillment of the conditions k_1=const, k_f=const or k_f/k_1=const. In case of larger k_1 (the equation

(35)) $\Sigma_i P_i^2 > 1/N$ [45]. For Witten-Sander clusters the value $\Sigma_i P_i^2$ is scaled as follows [45]:

$$\Sigma_i P_i^2 \sim N^{-\gamma}, \qquad (41)$$

where the exponent γ varies within the limits 0.5–0.8 [45].

One of the fractal analysis merits is a clear definition of limiting values of its main characteristics – fractal dimensions. So, the value d_{surf} changes within the limits $2 \pounds d_{surf} < 3$ [50]. At d_{surf}=2.0 the value N=0.1′10^5 relative units and according to the Figure. 17 plot maximal value S_c=1.65. At maximal for real solids dimension d_{surf}=2.95 [30] N=7.94′10^5 of relative units and according to Figure 17 plot $S_c \square 0$. This means that at such conditions trans-2-butene conversion degree goes to zero.

It was found out d_{surf} increase results to the decrease of trans-2-butene and cis-2-butene general conversion degree. This is explained by the fact that the formed in synthesis process polymeric chain has finite rigidity and consists of statistical segments of finite length. In virtue of this circumstance it can not "repeat" the growing catalyst surface roughness at d_{surf} increase and "perceive" it as still smoother surface.

In this case the effective fractal dimension of montmorillonite surface $d_{surf}^{\pounds}$ is determined as follows [26]:

$$d_{surf}^{ef} = d_{surf}, \qquad (42)$$

within the interval d_{surf}=2.0-2.5 and according to the equation (16) (at d_{surf}^{ef}=d_{surf} and

d_{surf}=d_{surf}^0) – within the interval 2.5-3.0.

As for the studied catalysts the values d_{surf}=2.637-2.776 (let's note their closeness

to the experimental value d_{surf}=2.78 [27]) then for d_{surf}^{ef} estimation the equation (16)

was used. In Fig. 18 the dependence $Q(d_{surf}^{ef})$ is adduced, which turns out to be linear

and is extrapolated to Q=0 at d_{surf}^{ef}=2.0 (or d_{surf}=3.0) and to Q=0 at d_{surf}^{ef}=d_{surf}=2.5.
Thus, the combined consideration of Figure 17 and 18 allows to assume the catalyst
optimal value d_{surf}, which is equal to 2.5. At this d_{surf} magnitude Q=1.0 and S_c=1.54, i.e.
close to maximal value S_c for montmorillonite in the considered reaction.

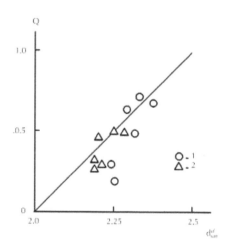

FIGURE 18   The dependence of general conversion degree Q on effective fractal dimension
d_{surf}^{ef} of catalyst surface for Na- montmorillonite (1) and Ca- montmorillonite (2) [46].

Hence, the results stated above demonstrated the important role of catalyst (montmorillonite) surface fractal geometry in its catalytic properties definition [46].

The polymerization *in situ* (together with a filler) is often applied as a method
for receiving nanocomposites. The matrix epoxy polymer cross-linking process both
in nanofiller (Na+- montmorillonite) presence and without it was studied within the
framework of a strange (anomalous) diffusion conception [51,52]. A nanocomposites
epoxy polymer/ Na+- montmorillonite (EP/MMT) cross-linking was made at temperatures 353, 373 and 393 K, cross-linking of EP – at 393 K [53].

Let us remind the main postulates of a strange (anomalous) diffusion conception.
As it is known [52], in general case diffusion processes are described according to the
equation:

$$\left\langle r^2 (t) \right\rangle \sim t^\alpha, \tag{43}$$

where $ár^2(t)ñ$ is mean-square displacement of a particle during time t.

If the exponent $\alpha = 1$ then the relationship (43) describes classical diffusion and if $\alpha^1 1$ – strange (anomalous) diffusion. Depending on concrete value α persistent (superdiffusive, $1 < \alpha £ 2$) and antipersistent (subdiffusive, $0 \square \alpha < 1$) processes are distinguished. In the strange diffusion equation the parameter α has significance of "active" time fractal dimension, in which real particles walks look as random process; active time interval is proportional to t^α [52]. In its turn, the exponent β in this equation [52]:

$$\frac{\alpha}{\beta} = \frac{d_s}{d} \tag{44}$$

Accounts for simultaneous particles jumps ("Levy flights") from one turbulence region into another. The ratio of exponents α/β can be determined according to the equation (44), in which spectral dimension d_s is determined, in its turn, according to the relationship (2).

The estimations have shown [51], that the introduction MMT in reactionary mixture increases substantially the value d_s: if for reaction without MMT $d_s = 0.027$, then at MMT introduction d_s increases up to 0.175–0.452 and raises at cross-linking temperature increase. A macromolecular coil (microgel) fractal dimension EP D_f can be determined with the help of the relationship (12) the account made according to the indicated ratio has shown, that the smallest value D_f equal to 1.52, is obtained for cross-linking reaction without MMT and at MMT introduction it increases up to 1.67–2.19.

The authors [51] have proceeded from the assumption, that the dependence Q on active time t^α should be linear and general for all the four studied reactions. The exponent β value was chosen on the base of this assumption. It turns out, that β decreases both at MMT introduction and at the temperature raising from 28 up to 3.8. Besides, at cross-linking reaction without MMT course the decrease β from 28 up to 9 is observed. In Figure 19 the dependence $Q(t^\alpha)$ for all the studied reactions is adduced. As one can see, the general linear correlation is obtained. The active time increases at MMT introduction: for reaction without MMT it makes $\sim 2\%$ from the real one and at MMT presence it can reach 30 %.

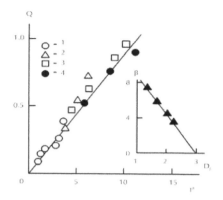

FIGURE 19   The Dependence of Cross-linking Degree Q on Active Time T^a for Reaction Without MMT (1) and in MMT presence at T=353 (2), 373 (3) and 393 K (4). In the insert: the dependence b(D_f) [51].

In the insert the dependence β on D_f is shown, from which β decrease at microgel fractal dimension growth follows. This assumes "Levy flights" probability decrease at the system viscosity increase. Hence, the ability to control active time gives the possibility of reaction course operation.

CONCLUSIONS

Thus, the present chapter results have been shown the applicability and usefullness of fractal analysis and strange (anomalous) diffusion conception for description of polymerization reactions, catalyzed by nanofillers. The nanofiller introduction in re-actionary mixture results in two-phase system formation, where a decisive role will be played by interfacial interactions. The polymerization conversion degree is defined by its active (fractal) time. Hence, the ability to control active time gives the possibility of reaction course operation.

KEYWORDS

- **Catalyst**
- **Fractal analysis**
- **Kinetics**
- **Nanofiller**
- **Polymerization**
- **Strange diffusion**

REFERENCES

1. Vasnev V.A., Naphadzokovz L.Kh., Tarasov A.I., Vinogradova S.V., Lependina O.L. Vysokomol. Soedin., 2000, vol. 42A(12), pp. 2065-2071. (Rus.).
2. Tyan H.-L., Liu Y.-C., Wei K.-H. Polymer, 1999, vol. 40(11), pp. 4877-4886.
3. Meakin P., Stanley H.E. J. Phys. A, 1984, vol. 17(1), pp. L173-L177.
4. Kozlov G.V., Shustov G.B. In book: Achievements in Polymer Physics-Chemistry Field. Ed. Zaikov G. a.a. Moscow, Khimiya, 2004, pp. 341-411. (Rus.).
5. Baranov V.G., Frenkel S.Ya., Brestkin Yu.V. Doklady AN SSSR, 1986, vol. 290(2), pp. 369-372. (Rus.).
6. Vilgis T.A. Physica A, 1988, vol. 153(2), pp. 341-354.
7. Rammal R., Toulouse G. J. Phys. Lett. (Paris), 1983, vol. 44(1), pp. L13-L22.
8. Naphadzokovz L.Kh., Kozlov G.V., Zaikov G.E. Teoreticheskie Osnovy Khimicheskoi Technologii, 2007, vol. 41(4), pp. 415-419 (Rus.).
9. Naphadzokovz L.Kh., Kozlov G.V., Zaikov G.E. J. Appl. Polymer Sci., 2007, vol. 105 (in press).
10. Kopelman R. In book: Fractals in Physics. Ed. Pietronero L., Tosatti E. Amsterdam, Oxford, New York, Tokyo, North-Holland, 1986, pp. 524-527.
11. Berstein V.A., Egorov V.M. Differential Scanning Calorimetry in Polymers Physics-Chemistry. Leningrad, Khimiya, 1990, 256 p. (Rus.).
12. Kozlov G.V., Novikov V.U. Uspekhi Fizichesk. Nauk, 2001, vol. 171(7), pp. 717-764. (Rus.).
13. Kozlov G.V., Zaikov G.E. Structure of the Polymer Amorphous State. Leiden-Boston, Brill Academic Publishers, 2004, 465 p.
14. Kozlov G.V., Dolbin I.V., Zaikov G.E. J. Appl. Polymer Sci., 2004, vol. 94(4), pp. 1353-1356.
15. Barns F.S. Biofizika, 1996, vol. 41(4), pp. 790-802. (Rus.).
16. Sahimi M., McKarnin M., Nordahl T., Tirrell M. Phys. Rev. A, 1985, vol. 32(1), pp. 590-595.
17. Naphadzokovz L.Kh., Malamatov A.Kh., Kozlov G.V. In coll.: The Scientific Proceedings of Young Scientists. Nal'chik, KBSU, 2006, pp. 305-307. (Rus.).
18. Novikov V.U., Kozlov G.V. Uspekhi Khimii, 2000, vol. 64(4), pp. 378-399. (Rus.).
19. Budtov V.P. Physical Chemistry of Polymer Solutions. Sankt-Peterburg, Khimiya, 1992, 384 p. (Rus.).
20. Temiraev K.B., Kozlov G.V., Sozaev V.A. Vestnik KBSU, Fizicheskie Nauki, 1988, No. 3, pp. 24-28. (Rus.).
21. Lipatov Yu.S. Interfacial Phenomena in Polymers. Kiev, Naukova Dumka, 1980, 260 p. (Rus.).
22. Pfeifer P. In book: Fractals in Physics. Ed. Pietronero L., Tosatti E. Amsterdam, Oxford, New York, Tokyo, North-Holland, 1986, pp. 72-81.
23. Kozlov G.V., Zaikov G.E., Lipatov Yu.S. Doklady NAN Ukraine, 2002, No. 8, pp. 130-135. (Rus.).
24. Naphadzokovz L.Kh., Kozlov G.V., Zaikov G.E. Plast. Massy, 2008, (in press) (rus.).
25. Kozlov G.V., Shustov G.B. Khimicheskaya Technologiya, 2006, No. 1, pp. 24-26. (Rus.).
26. Van Damme H., Levitz P., Bergaya F., Alcover J.F., Gatineau L., Fripiat J.J. J. Chem. Phys., 1986, vol. 85(1), pp. 616-625. 27.   Pernyeszi T., Dekany I. Colloid Polymer Sci., 2003, vol. 281(1), pp. 73-78.
28. Naphadzokovz L.Kh., Kozlov G.V., Tlenkopachev M.A. 3-rd Intern. Symposium on Hybridized Materials with Super-Functions. Monterrey, Mexico, 3-6th December 2006, PM-06, p. 75.

19.   29. Naphadzokovz L.Kh., Kozlov G.V., Zaikov G.E. In book: Chemical and Biochemical
      Physics, Kinetics and Thermodynamics. New Pespectives. Ed. Scott P., Zaikov G., Kablov
      V. New York, Nova Science Publishers, Inc., 2007, (in press) (Rus.).
20.   30. Balankin A.S. Synergetics of Deformable Body, Moscow, Ministry of Defence SSSR,
      1991, 404 p. (Rus.).
21.   31. Kozlov G.V., Sanditov D.S., Lipatov Yu.S. In book: A chievements in Polymer Phys-
      ics-Chemistry Field. Ed. Zaikov G. a.a. Moscow, Khimiya, 2004, pp. 412-474. (Rus.).
22.   32. Sanditov D.S., Bartenev G.M. Physical Properties of Disordered Structures. Novosi-
      birsk, Nauka, 1982, 256 p. (Rus.).
23.   33. Kozlov G.V., Aloev V.Z. Percolation Theory in Polymers Physics-Chemistry.
      Nal'chik, Polygraphservic I T, 2006, 148 p. (Rus.).
24.   34. Bobryshev A.N., Kozomazov V.N., Babin L.O., Solomatov V.I. Synergetics of Com-
      posite Materials. Lipetsk, NPO ORIUS, 1994, 154 p. (Rus.).
25.   35. Kozlov G.V., Naphadzokovz L.Kh., Zaikov G.E. Prikladnaya Fizika, 2007, No. 4, pp.
      52-55. (Rus.).
26.   36. Kozlov G.V., Zaikov G.E. J. Balkan Tribologic. Assoc., 2004, vol. 10(1), pp. 1-30.
27.   37. Naphadzokovz L.Kh., Kozlov G.V., Zaikov G.E. Electron. Zhurnal "Studied in Rus-
      sia", 111, p. 1173-1179, 2007, http: // Zhurnal. apl. relarn. ru / articles / 2007 / 111. pdf.
      (Rus.).
28.   38. Brady R.M., Ball R.C. Nature, 1984, vol. 309(5965), pp. 225-229.
29.   39. Botet R., Jullien R., Kolb M. Phys. Rev. A, 1984, vol. 30(4), pp. 2150-2152.
30.   40. Kobayashi M., Yoshioka T., Imai M., Itoh Y. Macromolecules, 1995, vol. 28(22), pp.
      7376-7385.
31.   41. Kozlov G.V., Batyrova H.M., Zaikov G.E. J Appl. Polymer Sci., 2003, vol. 89(7), pp.
      1764-1767.
32.   42. Nugmatullin R.R. Teoretich. i Matematich. Fizika, 1992, vol. 90(3), pp. 354-367.
      (Rus.).
33.   43. Kozlov G.V., Shustov G.B. In coll.: International Interdisciplinary Seminar "Frac-
      tals and Applied Synergetics, FaAS-01", Moscow, Publishers MSOU, 2001, pp. 155-157.
      (Rus.).
34.   44. Melikhov I.V. Rossiiskii Khimichesk. Zhurnal, 2002, vol. 66(5), pp. 7-14. (Rus.).
35.   45. Meakin P. Chem. Phys. Lett., 1986, vol. 123(5), pp. 428-432.
36.   46. Naphadzokovz L.Kh., Shustov G.B., Kozlov G.V. In cool.: Proceedings of 9-th Inter-
      national Symposium "Order, Disorder and Properties of Oxides, ODPO-2006". Rostov-
      na-Donu, RSU, 2006, pp. 49-51. (Rus.).
37.   47. Moronta A., Ferrer V., Quero J., Arteaga G., Choren E. Appl. Catalysis A, 2002, vol.
      230(1), pp. 127-135.
38.   48. Stanley E.H. In book: Fractals in Physics. Ed. Pietronero L., Tosatti E. Amsterdam,
      Oxford, New York, Tokyo, North-Holland, 1986, pp. 463-477.
39.   49. Naphadzokovz L.Kh., Aphashagova Z.Kh., Kozlov G.V. In coll.: V Intern. Sci.-Pract.
      Conf. "Effective Building Constructions: Theory and Practice". Penza, PSU, 2006, pp.
      271-273. (Rus.).
40.   50. Avnir D., Farin D., Pfeifer P. Nature, 1984, vol. 308(5959), pp. 261-263.
41.   51. Naphadzokovz L.Kh., Kozlov G.V. In coll.: Intern. Sci.-Techn. Conf. "Composite
      Building Materials. Theory and Practice". Penza, PSU, 2006, pp. 170-172. (Rus.).
42.   52. Zelenyi L.M., Milovanov A.V. Uspekhi Fizichesk. Nauk, 2004, vol. 174(8), pp. 809-
      852. (Rus.).
43.   53. Chen J.-S., Poliks M.D., Ober C.K., Zang Y., Wiesher U., Giannelis E. Polymer, 2002,
      vol. 43(18), pp. 4895-4904.

CHAPTER 5

SOLUBLE BLOCK COPOLYIMIDES

N. I. LOBA, V. M. ZELENCOVSKY, N. R. PROKOPCHUK,
and E. T. KRUTKO

CONTENTS

ABSTRACT

Fragmentary polyamicacids on the basis of industrially made poly(4,4'-diaminodipheniloxide)piromellitamicasid and oligoamidoacid, received by polycondensation of 4,4'-diaminodiphenyloxide and dianhydride 4,8-diphenil-1,5-diazodicyclic-/3,3,0/-octane-2,3,6,7-tetracarboxylic acid has been synthesized with their subsequent chemical imidization. Unlike initial poly(4,4'-dipheniloxide)peromellitimide synthesized block polyimides possess solubility in polar aprotic solvents. For explanation of their solubility calculations of parameters of conformation of constitutional repeating units of macromolecules by computer modeling has been carried out.

5.1   INTRODUCTION

One of the current trends in the polyimides' synthesis is the creation of fusible and soluble in organic solvents materials [1-7]. This allows to extend the range of their practical use and refines the classical methods for thermoplastics. This problem is particularly relevant in cases when the traditional scheme of hightemperature prepolymer conversion into the final polymer, which usually takes place in the final product, can not be carried out due to thermal instability of the product elements. It is often achieved by the use of monomers (diamines and dianhydrides) with bulky side groups for the synthesis of polyimides [8-12].

The synthesis of polymers from a mixture of several diamines and dianhydrides, and especially the synthesis of block copolyimides, represents wide opportunities of directed regulation of polyimides properties, including giving them solubility. One of the methods of the block copolymers synthesis is getting them on the basis of pre-synthesized oligomers with determination molecular weight and with different functional groups.

In this chapter the synthesis of poly (4,4'-dipheniloxide)pyromellit(amic acid) (PAA), fragmented by oligo(amic acid) (OAA), obtained by low-temperature polycondensation of 4,4'-diaminodiphenyl oxide and dianhydride 4,4'-diphenyl-1,5-diazobicyclo/3,3,0/octane-2,3,6,7-tetracarboxilic acid and its subsequent chemical imidization. In the opposition to the original poly-(4,4'-aminodiphenyl)-pyromellitimide (PI) synthesized block copolyimides (BSPI) have a solubility in polar aprotonic solvents. The parameter of BSPI conformation was calculated to explain its solubility.

5.2   SUBJECT AND METHODS OF RESEARCH

The synthesis of fragmented poly(amic acids) was implemented by adding into 13% solution of PAA in dimethylformamide (DMFA) OAA as a dry substance in quantities of 3, 4, 5, 7, 9, 12, 15, 20, 25 wt. % of the content of dry PAA in solution. The relative viscosity of the 0.5% solution of PAA in DMFA was 1,87, similar characteristic of OAA was 1,23.

The interaction is described by the following reaction:

The transformation of the synthesized block copolymers which consists of frag-ments of PAA (A) and OAA fragments (B) into polyimide was realized by two ways – chemical and thermal [13]. The research of the spectral characteristics of the synthe-sized block copolyimide films was performed on FTIR spectrometer Nicolet 7101 in the frequency range 4000–300 cm^{-1}.

For defining the geometric parameters of the polyimide macromolecules frag-ments semi-empirical quantum-chemical calculations were carried out using Hamilto-nian PM6 model [14] with software package MOPAC 2009 . All geometric parameters of the three systems, including two elementary level of poly(4,4'-diphenyloxide) pi-romellitimide (PI), oligoimide based on 4,4'-diaminodiphenyloxide and dianhydride 4,8-diphenyl-1,5-diazobicyclo/3,3,0/octane-2,3,6,7-tetracarboxilic acid (OI) and co-polyimide (CPI), which contains links of PI and OI were fully optimized in order to determine the structural characteristics. According to the calculated geometric param-eters the values of the identity period C, and contour length of the fragment of the macromolecular chains in the range of a period of identity L were determined . Figure 1 shows the plane projection of the synthesized block copolymer fragments based on PAA and the OAA and illustrates the definition of L and C.

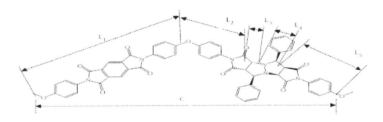

FIGURE 1 The scheme of block copolyimide fragments to determine its structural characteristics

The conformation parameter K_n was calculated to quantify the crookedness and curl of different nature soluble block copolyimides fragments [15-19]:

$$K_\pi = \frac{L-C}{C} \cdot 100\% \quad ; \quad L = \sum_{i=1}^{n} L_i$$

where C – identity period along the polymer chain, Å;
L – contour length of the chain skeleton within a period of identity, Å.

5.3   RESULTS AND DISCUSSION

It was experimentally established that synthesized OAA well combine with a solution of aromatic PAA and can be introduced into it by dissolving by stirring for 2–2,5 hours with up to 20 wt. % from the content of aromatic PAA (in terms of dry matter) to the form of a homogeneous clear solution in DMFA. However, film-forming ability of these compositions only retains if the content of the input oligomer PAA is less than 9 wt. %. Exceeding of these limit results leads to bursting and loosing the ability to form a uniform film by the fragmented polymer after removal of the solvent by drying in air at room or elevated temperature to 50–60 °C. Lower limit of the content of the OAA, ensuring the achievement of the target properties of the composition, is 4 wt. %. Its fewer content in substance does not provide solubility to chemically imidizied compositions.

It was studied the kinetics of transformation macromolecules of synthesized BSPI prepolymers by IR spectroscopy [13] in order to optimize the process of obtaining fragmented polyimides. The comparison of the intensity of the absorption bands of imide rings with a peak at 1380 cm⁻¹ before and after the chemical imidization showed that the maximum degree of imidization is 92–93% and it is reached within 60 min. In a similar way it was established that after thermal imidization by step increase of temperature up to 320 °C within 120 min maximum degree of imidization of the block copolyimides is 86–87%.

It is found that the rate constant for the initial stage of the chemical imidization of blok copoly(amic acids) measured at temperatures of 22, 35 and 45 °C, are about two times higher than the rate constants of this stage of the thermal imidization, measured at 170, 185 and 205 °C. In this case, the activation energy of the chemical imidization is 72±5 kJ/mol, and for the process of thermal imidization is 87±5 kJ/mol, which is according to the literature data [13]. These pecularities of the process, probably due to the fact that the mobility of the carboxyamide links in the liquid phase of the chemical imidization are much higher than in the condensed phase at a solid thermal imidization.

It should be emphasized that the synthesized chemically imidizied blok copolyimide in contrast to a similar polymer obtained by thermal imidization, soluble in polar aprotic solvents – dimethylformamide, dimethylacetamide, dimethylsulfoxide, N-methylpyrrolidone, forming a normal solution concentration up to 5 wt% from which coatings with high electrophysical characteristics can be obtained.

Apparently this is due to the fact that under mild conditions of the chemical imidization flowing in solution under the influence of dehydrating reagent mixture in the polymer intermolecular interactions difficulted, leading to the loss of polymer solubility. Oligoamic acids fragments containing voluminous diphenyldiazobeziclooktan links create difficulties for the intermolecular interactions.

It is of a great interest to determine the causes of acquisition of block copolyimides solubility. In this respect the analysis of the structural and conformational properties of the polymer dives is useful information.

In real materials the individual molecules are not isolated from each other. There are always certain interchain interaction between them, whose energy depends on the chemical nature of the parts of the macromolecules and the density of packing. The distance between the interacting atoms and atomic groups in neighboring macromolecules depends on the conformation of the macromolecular chains. It is shown on the example of aromatic and cycloaliphatic polyimides [6,8] that there is a clear correlation between the energy of intermolecular interactions and conformational parameter K_p. Conformational parameter K_p reduces in the transition from the saddle geometric structure of cycloaliphatic dianhydride to plane geometry of piromellitic dianhydride, and from the broken structure of "flexible" diamines to a flat linear structure of "hard" diamines, packing density of the macromolecules and intermolecular interaction energy are increases. That is one of the causes of solubility of aromatic polyimides.

Table 1 shows the three-dimensional fragments of elementary components of PI, OI, and CPI and shows the values of the identity periods (C, Å), contour length of the chains within a period of identity (L, Å), and the conformational parameters (K_p).

TABLE 1 Conformations and structural characteristics of three-dimensional block copolyimides fragments

Fragment of macromolecule	L, Å	C, Å	K_p, %
PI	36,014	32,752	9,9
OI	39,472	24,150	63,0

TABLE 1 *(Continued)*

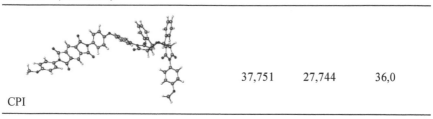 CPI	37,751	27,744	36,0

The data presented in the table show that the conformational parameter is minimal for elementary units of PI based on pyromellitic dianhydride and 4,4'-diaminodiphe-niloxside because bending of the polymer chain is only in the "hinge" oxygen atom diamine fragment of the repeating unit of the macromolecule. The elementary unit of OI has a much higher value of conformation, because of the collapse fragment chain macromolecules as bending occurs not only in the "hinge" oxygen, but in fragments diazobicyclooctan dianhydride acid. In elementary links of CPI with fragments of PI and OI, due to the combination of rigid fragment of PI and OI strongly curved fragment conformational parameter takes an intermediate value, which, however, is 3,6 times higher the value of the IP for the aromatic fragment (PI). High values of the conformational parameters and hence the low level of intermolecular interactions helps to create a more "loose" structure, the penetration of the solvent molecules between the polymer chains, their salvation, swelling and dissolution.

5.4   CONCLUSION

Thus, the results of the research indicate that it is possible to use oligoamic acid based on 4,4'-diaminodipheniloxside and dianhydride 4,4'-diphenyl-1,5-diazobicyclo/3,3,0/ octane-2,3,6,7-tetracarboxilic acid which have a volumetric three-dimensional structure and block the implementation of interactions between aromatic PAA macromolecules as copolymer. We can get polyimide film-forming materials with high dielectric properties, which are soluble in aprotic amide solvents by chemical imidization.

KEYWORDS

- **Block copolyimides**
- **Conformation**
- **Copolyimides**
- **Geometric parameters**
- **Identity period**
- **Oligo(amic acid)**
- **Polyimides**
- **Structural characteristics**

REFERENCES

1. Clemenson, P. I. et al. Synthesis and characterization of new water-soluble precursors of polyimides. *Polym. Eng. Sci.* **37**(6), 966–977 (1997).
2. Smirnova, V. Influence of planar molecular orientation on mechanical properties of hard-links polyimide films. *Vysocomol. Soedin. Ser. A.*, **49**(10), 1810–1816 (2007).
3. Vidyakin, M., Lazareva, J., and Yampolsky, Y. *Vysocomol. Soedin. Ser. A.*, **49**(1), 1703–1711 (2007).
4. Chern, Y. T. Low dielectric constants of soluble polyimides derived from the novel 4,9-bis(4-(4-aminophenoxy)-phenyl)-diadamantane Y. T. Chen and H. Ch. Shiue (Eds.) *Macromolecules*, **30**(19), 5766–5772 (1997).
5. Liou, G. S. et al. Preparation and properties of new soluble aromatic polyimides from 2,2'-bis(3,4-dicarboxyphenoxy) biphenyl dianhydride and aromatic diamines. *J. Polym. Sci. Part A, Polym. Chem.* **36**(12), 2021–2027 (1998).
6. Imai, Y. Synthesis of novel organic-soluble high-temperature aromatic polymers. *High Perfom. Polym.* **7**(3), 337–345 (1995).
7. Imai, Y. Resent progress in synthesis of polyimides. *J. Photopolym. Sci. Tecnol.* **7**(2), 251–256 (1994).
8. Yang, C. P. et al. Chemical modification of 3,3',4,4'-biphenyltetracarboxylic dianhydride polyimides by a catechol-derived bis(ether amine). *J. Appl. Polym. Sci.* **84**(2), 351–358 (2002).
9. Babman, T. Preparation and properties of novel polyimides derived from 4-aril-2,6-bis(4-aminophenyl)pyridine. T. Babman and Y. Hamid (Eds). *J. Polym. Sci. Part. A, Polym. Chem.* **39**(21), 3826–3831 (2001).
10. Lee, D. H. et al. Preparation and properties of aromatic polyimides based on 4,4 -(2,2,2-trifluoro-1-phenylethylidene)diphthalic anhydride. *J. Appl. Polym. Sci.* **85**(1), 38–44 (2002).
11. Zhou, H. et al. Soluble fluorinated polyimides derived from (4'-aminophenoxy)-2-(3'-trifluoromethylphenyl)-benzene and aromatic dianhydrides. *J. Polym. Sci. Part A, Polym. Chem.* **39**(14), 2404–2413 (2001).
12. Reddy, D. S. Synthesis and characterization of soluble polyimides derived from 2,2'-bis(3,4-dicarboxyphenoxy)-9,9'-spirobifluorene dianhydride. D. S. Reddy, C. F. Shu, and F. I. Wu (Eds). *J. Polym. Sci. Part A, Polym. Chem.* **40**(2), 262–268 (2002).
13. Laius, L. A., and Tsapovetsky, M. I. In Polyamic Acids and Polyimides. Transformation of solid polyamic acids at thermal treatment. Bessonov. M. I. Zubkov and V. A. Boca Raton, Fla (Eds). p. 310 (1993).
14. Stewart, J. J. P. Optimization of Parameters for Semiempirical Methods V. Modification of MDDO Aproximations and Application to 70 Elements. J. J. P. Stewart, and J. Mol. Mod. (Eds) **13**, 1173–1213 (2007).
15. Korshavin, L. N., Prokopchuk, N. R., Sidorovich, A. V., Milevskaya, I. S., Baclagina, Y. G., and Koton, M. M. Correlation of chains configuration, structure and mechanical properties of polypiromellitimide fibers // *Reports of the Academy of Sciences of the USSR.* **236**(1), 127–130 (1977).
16. Prokopchuk, N. R., Korjavin, L. N., Backlapsha, Y. G., Frenkel, S. Y., and Koton, M. M. The relationship of the chemical structure and mechanical properties of oriented polyarilenimides. *Vysocomol. Soedin. Ser. A.*, **18**(3), 707–712 (1976).
17. Koton, M. M., Prokopchuk, N. R., Florinsky, F. S., Frenkel, S. Y., Korshavin, hL. N., and Pushkina, T. P. Procede de fabrication de fibers de polyimide. Paris France Depot № 7823262. 1981.

18. Koton, M. M., Prokopchuk, N. R., Florinsky, F. S., Frenkel, S. Y., Korshavin, hL. N., and Pushkina,, T. P. Method for obtaining polyimide fibres. London UK Patent GB 202531 IB. 1982.
19. Koton, M. M., Prokopchuk, N. R., Florinsky, F. S., Frenkel, S. Y., Korshavin, hL. N., and Pushkina, T. P. Tokyo. Method for obtaining polyimide fibres. Japan Patent № 1147596. 1983.

CHAPTER 6

HYALURONAN DEGRADATION UNDER FREE-RADICAL OXIDATION STRESS: ACTION AND HEALING

T. M. TAMER

CONTENTS

6.1 INTRODUCTION

Oxidation stress is unbalanced between prooxidants and natural antioxidants in body that lead to several diseases such as rheumatoid. Hyaluronic acid (HA), is a high molecular weight biopolysacharide, is found in the extracellular matrix of soft connective tissues and is particularly concentrated in synovial fluid (SF). Half-live time of Hyaluronan in SF is approximately 12 hrs in normal conditions. This process is accelerated under normal oxidation stress that generates troubles in human joints.

This chapter describe oxidation stress-source and effects-Hyaluronan origin, properties and functions, and finally thiol compounds as antioxidants preventing HA degradations under conditions of oxidation stress.

The ability to utilize oxygen has provided humans with the benefit of metabolizing lipids, proteins, and carbohydrates for energy; however, it does not come without cost. Oxygen is a highly reactive atom that is capable of becoming part of potentially damaging molecules commonly called "free radicals." Free radicals are capable of attacking the healthy cells of the body, causing them to lose their structure and function. The cell damage caused by free radicals appears to be a major contributor to aging and to degenerative diseases of aging such as cancer, cardiovascular disease, cataracts, immune system decline, and brain dysfunction. [1]. Overall, free radicals have been implicated in the pathogenesis of at least 50 diseases. [2,3]

Fortunately, free radical formation is controlled naturally by various beneficial compounds known as antioxidants. It is when the availability of antioxidants is limited that this damage can become cumulative and debilitating.

Oxidative stress is the phenomenon that occurs when the steady-state balance of pro-oxidants to antioxidants is shifted in the direction of the former, creating the potential for organic damage. Pro-oxidants are by definition free radicals, atoms or clusters of atoms with a single unpaired electron. Physiologic concentrations of pro-oxidants are determined both by internal and external factors. Pro-oxidant reactive oxygen species (ROS), for example, are normal products of aerobic metabolism. However, under pathological conditions ROS production can increase, surpassing the body's detoxification capacity and thus contribute to molecular-level organic pathology. External sources of free radicals include exposures to environmental toxins such as ionizing radiation, ozone and nitrous oxide, cigarette smoke (including passive inhalation) and heavy metals, as well as dietary intake of excess alcohol, unsaturated lipid, and other chemicals and compounds present in food and water.

Antioxidants are chemical compounds that can bind to free radicals and thus prevent them from damaging healthy cells. Antioxidants can be divided into enzymatic and non-enzymatic subtypes. Several antioxidant enzymes are produced by the body, with the three major classes being catalase, the glutathione (GSH) peroxidases, and the superoxide dismutases (SODs). Non-enzymatic antioxidants include the innate compound glutathione as well as antioxidant vitamins obtained through the diet, such as α-tocopherol (vitamin E), ascorbic acid (vitamin C), and β-carotene.

6.2   FREE RADICALS DEFINITION AND FORMATION

Free radicals are electrically charged molecules, i.e., they have an unpaired electron, which causes them to seek out and capture electrons from other substances in order to neutralize themselves.

Although the initial attack causes the free radical to become neutralized, another free radical is formed in the process, causing a chain reaction to occur. And until subsequent free radicals are deactivated, thousands of free radical reactions can occur within seconds of the initial reaction.

6.2.1   SOURCE OF FREE RADICAL

In cells, there are two main sources of superoxide anion (O_2^-) and hydrogen peroxide (H_2O_2). Hydroxyl radical (HO) is generated from superoxide anion (O_2^-) and hydrogen peroxide (H_2O_2).

6.2.2   SOURCES OF SUPEROXIDE RADICAL (O_2^-)

The following Table (1) illustrates the most important reactions within the cell that generate superoxide anion (O_2^-).

TABLE 1   Reaction sources of superoxide anion radical (O_2^-)

Source	Pathophysiological significance
• *Enzymic reactions*	
- xanthine oxidase	Intestinal ischemia/reperfusion
- NADH oxidase	Present in leukocytes: bactericidal activity
- NADPH-cytochrome P450 reductase	
•*Cellular sources*	
- leukocytes and macrophages	Bactericidal activity
- mitochondrial electron transfer	
- microsomal monooxygenase	
• *Environmental factors*	
- ultraviolet light	
- X rays	
- toxic chemicals	
- aromatic hydroxylamines	

- aromatic nitro compounds

- insecticides, such as paraquat

- chemotherapeutic agents, such as quinines

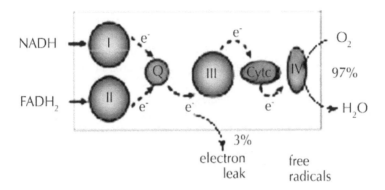

Mitochondria are major cellular sources of reactive oxygen species. Mitochondria consume oxygen associated with the process of oxidative phosphorylation. Under normal conditions, approximately 95–97% of the oxygen is reduced to water; electron leakage, accounting for about 3–5% of the total oxygen consumed by mitochondria, is associated with the generation of oxygen radicals.

6.2.3 SOURCES OF HYDROGEN PEROXIDE

Hydrogen peroxide (H_2O_2) is generated within the cells by two distinct processes: 1) nonradical or enzymic generation and 2) radical or from superoxide anion disproportionation.

NONRADICAL OR ENZYMIC GENERATION

The following enzymes do generate (H_2O_2) upon reduction of their cosubstrate, molecular oxygen:

Glycolate oxidase, d-amino acid oxidase, urate oxidase, acetyl-CoA oxidase, NADH oxidase and monoamine oxidase.

The latter enzyme, monoamine oxidase (MAO) occurs in two forms A and B and it catalyzes the oxidative deamination of biogenic amines. It is present in the outer mitochondrial membrane.

RADICAL GENERATION FROM SUPEROXIDE ANION DISPROPORTIONATION

This is achieved upon dismutation or disproportionation of superoxide anion (O_2^-), according to the reaction mentioned before:

$$O_2^- + O_2^- + H^+ \rightarrow H_2O_2 + O_2$$

As mentioned above, *mitochondria are major cellular sources of oxyradicals*. Superoxide anion radical (O_2^-), generated upon autoxidation of ubisemiquinone, is vectorially released into the inter membrane space and the mitochondrial matrix. In the latter compartment, O_2^- dismutates to H_2O_2.

The H_2O_2 is a freely diffusible species that can cross membranes. Hence, mitochondria have two major sources of H_2O_2: On the one hand, H_2O_2 generated by disproportionation of superoxide anion in the mitochondrial membrane and, on the other hand, the oxidative deamination of biogenic amines by the outer mitochondrial membrane-bound monoamine oxidase activity. Mitochondrion-generated H_2O_2 is involved in the redox regulation of cell signaling pathways. The steady-state levels of H_2O_2 ($[H_2O_2]ss$) determine the cellular redox status and the transition from proliferation to apoptosis and necrosis.

6.2.4 SOURCES OF HYDROXYL RADICAL

The most of the hydroxyl radical (HO·) generated *in vivo*, except for that during excessive exposure to ionizing radiation, originates from the breakdown of hydrogen peroxide (H_2O_2) *via* a *Fenton reaction*.

The Fenton reaction entails a metal-dependent reduction of hydrogen peroxide (H_2O_2) to hydroxyl radical (HO). *Transition metals*, such as copper (Cu), iron (Fe), and cobalt (Co), in their reduced form catalyze this reaction:

$$Fe^{2+} + H_2O_2 \rightarrow Fe^{3+} + HO^- + HO·$$

As indicated above, the Fenton reaction requires the transition metal in its reduced state. Reduction of the transition metal may be accomplished by superoxide anion (O_2^-), as in the following example with Fe^{+++}:

$$Fe^{+++} + O_2^- \rightarrow Fe^{++} + O_2$$

The overall reaction, involving iron reduction by superoxide anion (O_2^-) and iron oxidation by hydrogen peroxide (H_2O_2), is as follows:

$$Fe^{3+} + O_2^- \rightarrow Fe^{2+} + O_2$$
$$Fe^{++} + H_2O_2 \rightarrow Fe^{+++} + HO^- + HO·$$

$$O_2^- + H_2O_2 \rightarrow O_2 + HO^- + HO\cdot$$

The latter reaction ($O_2^- + H_2O_2 \rightarrow O_2 + HO^- + HO\cdot$), is known as the *Haber-Weiss reaction*. This reaction, as such, proceeds at very slow rates. The *Fenton reaction*, that is, metal catalyzed reduction of hydrogen peroxide (H_2O_2), prevails in a biological environment. It is worth noting that, at variance with superoxide anion (O_2^-) and hydrogen peroxide (H_2O_2), there is no direct generation of hydroxyl radical (HO) in the cell. Both, superoxide anion and hydrogen peroxide are required to form the highly reactive hydroxyl radical (HO).

6.2.5 SINGLET OXYGEN

Singlet oxygen is a reactive oxygen species that can be formed not only by energy transfer (as mentioned above), but also by electron-transfer reactions.

Electron transfer reactions: Of biological interest, the enzyme *myeloperoxidase*, present in neutrophiles, can catalyze the formation of hypochlorite from Cl^- and H_2O_2. The further reaction of hydrogen peroxide (H_2O_2) with formed HOCl yields singlet oxygen ($^1O_2$):

$$Cl^- + H_2O_2 \xrightarrow{\textit{myeloperoxidase}} HOCl + H_2O$$

$$HOCl + H_2O_2 \rightarrow Cl^- + H_2O + H^+ + {}^1O_2$$

Energy transfer reactions: This is another way to generate singlet oxygen as it is comprised in the photosensitization of different chemotherapeutic agents. The chemotherapeutic agent (or sensitizer = S) absorbs energy upon irradiation and transfers this energy to molecular oxygen with formation of singlet oxygen ($^1O_2$).

$$S + h\upsilon \rightarrow S^*$$

$$S^* + O_2 \rightarrow S + {}^1O_2$$

As mentioned before, singlet oxygen ($^1O_2$) is a reactive species that reacts with molecules, such as vitamin E, vitamin C, DNA, cholesterol, carotenoids, polyunsaturated fatty acids in membranes, and certain amino acids.

HARMFUL OF FREE RADICAL

All the biological molecules present in our body are at risk of being attacked by free radicals. Such damaged molecules can impair cell functions and even lead to cell death eventually resulting in diseased states.

It can be considered that superoxide anion radical (O_2^-) and hydrogen peroxide (H_2O_2) are less reactive than hydroxyl radical $(HO^{\cdot})$ and singlet oxygen. However, in a suitable biological setting the two first species may display considerable chemical reactivity leading to damage of various biomolecules

LIPIDS AND LIPID PEROXIDATION

Membrane lipids present in subcellular organelles are highly susceptible to free radical damage. Lipids when reacted with free radicals can undergo the highly damaging chain reaction of lipid peroxidation (LP) leading to both direct and indirect effects. During LP a large number of toxic byproducts are also formed that can have effects at a site away from the area of generation, behaving as 'second messengers'. The damage caused by LP is highly detrimental to the functioning of the cell (Ramsarma T et al., 2003).

Lipid peroxidation (LP) is a free radical mediated process. Initiation of a peroxidative sequence is due to the attack by any species, which can abstract a hydrogen atom from a methylene group (CH_2), leaving behind an unpaired electron on the carbon atom (CH). The resultant carbon radical is stabilized by molecular rearrangement to produce a conjugated diene, which then can react with an oxygen molecule to give a lipid peroxyl radical (LOO) These radicals can further abstract hydrogen atoms from other lipid molecules to form lipid hydroperoxides (LOOH) and at the same time propagate LP further. The peroxidation reaction can be terminated by a number of reactions. The major one involves the reaction of $LOO^{\cdot}$ or lipid radical $(L^{\cdot})$ with a molecule of antioxidant such as vitamin E or α-tocopherol (α-TOH) forming more stable phenoxyl radical that is not involved in further chain reactions. This can be 'recycled' by other cellular antioxidants such as vitamin C or GSH.

$$LH^{\cdot} + OH \rightarrow L^{\cdot} + H_2O$$

$$L^{\cdot} + O_2 \rightarrow LOO$$

$$LOO^{\cdot} + LH \rightarrow L^{\cdot} + LOOH$$

$$LOO^{\cdot} + \alpha\text{-TOH} \rightarrow LOOH + \alpha\text{-TO}^{\bullet}$$

The process of LP, gives rise to many products of toxicological interest like malondialdehyde (MDA), 4-hydroxynonenal (4-HNE) and various 2-alkenals. Isoprostanes are unique products of lipid peroxidation of arachidonic acid and recently tests such as mass spectrometry and ELISA-assay kits are available to detect isoprostanes [4].

6.2.6 CARBOHYDRATES

Free radicals such as ˙OH react with carbohydrates by randomly abstracting a hydrogen atom from one of the carbon atoms, producing a carbon-centered radical. This leads to chain breaks in important molecules like hyaluronic acid. In the synovial fluid surrounding joints, an accumulation and activation of neutrophils during inflammation produces significant amounts of oxyradicals that is also being implicated in rheumatoid arthritis.

6.2.7 DNA OXIDATION

Hydroxyl radical is endowed with unique properties: due to a combination of high electrophilicity, high thermochemical reactivity, and a mode of production that can occur in the vicinity of DNA (site specific mechanism), it can both abstract H atoms from the sugar in the DNA helix and add to DNA bases, leading to single strand breaks and nucleobase (8- hydroxydesoxyguanosine) oxidation, respectively.

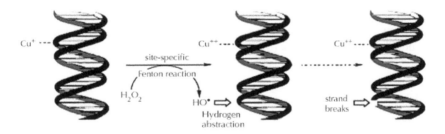

FIGURE 1   Hydrogen abstraction–DNA strand breaks.

ADDITION - NUCLEOBASE OXIDATION

Hydroxyl radical addition to bases such as guanine, proceeds very rapidly and leads to the formation of 8-hydroxydesoxyguanosine, which is used as a fingerprint of nucleobase oxidative damage.

desoxy guanosine   8-hydroxydesoxyguanosine

The DNA is susceptible to changes that would lead to mutations. For example, DNA bases are damaged by an encounter with free radicals or environmental chemicals. Hydroxyl radical mediated damage on sugars (deoxyribose) is a part of the known C'4 mechanism and leads to strand breaks. Oxidative damage of bases, usually leads to adduct formation, as exemplified above with 8-hydroxydesoxyguanosine.

Reactive oxygen species can damage DNA at different levels: hydroxyl radical through addition reactions can cause single-strand base damage (e.g., formation of 8-hydroxyldeoxyguanosine) and through H abstraction single strand DNA nick (ssDNA nick) or double strand DNA break (dsDNA break).

Oxidative damage to DNA is a result of interaction of DNA with ROS or RNS. Free radicals such as OH and H˙ react with DNA by addition to bases or abstractions of hydrogen atoms from the sugar moiety. The C4–C5 double bond of pyrimidine is particularly sensitive to attack by OH, generating a spectrum of oxidative pyrimidine damage products, including thymine glycol, uracil glycol, urea residue, 5-hydroxy-deoxyuridine, 5-hydroxydeoxycytidine, hydantoin and others. Similarly, interaction of ˙OH with purines will generate 8-hydroxydeoxyguanosine (8-OHdG), 8-hydroxy-deoxyadenosine, formamidopyrimidines and other less characterized purine oxidative products. Several repair pathways repair DNA damage (Halliwell B and Aruoma OI., 1993). 8-OHdG has been implicated in carcinogenesis and is considered a reliable marker for oxidative DNA damage.

6.2.8 PROTEINS

Oxidation of proteins by ROS/RNS can generate a range of stable as well as reactive products such as protein hydroperoxides that can generate additional radicals particularly upon interaction with transition metal ions. Although most oxidized proteins that are functionally inactive are rapidly removed, some can gradually accumulate with time and thereby contribute to the damage associated with ageing as well as various diseases. Lipofuscin, an aggregate of peroxidized lipids and proteins accumulates in lysosomes of aged cells and brain cells of patients with Alzheimer's disease [5].

6.2.9 METHODS TO OVERCOME FREE RADICAL RISKS

To protect the cells and organ systems of the body against reactive oxygen species, humans have evolved a highly sophisticated and complex antioxidant protection system. It involves a variety of components, both endogenous and exogenous in origin, that function interactively and synergistically to neutralize free radicals[6].

In their definition of the term antioxidant, Halliwell and Gutteridge (1989) state, 'any substance that, when present at low concentrations compared to that of an oxidizable substrate, significantly delays or inhibits oxidation of that substrate'. This definition would comprise compounds of nonenzymic as well as enzymic nature. Table 2 overviews some of the antioxidants of biological interest [7].

TABLE 2   Antioxidant defense in biological systems. Condensed list of antioxidant compounds and enzymes. Modified from (Sies, H., 1985)

System	Remarks
Non-enzymic	
α-Tocopherol (vitamin E)	radical chain-breaking
ß-Carotene	singlet oxygen quencher
Lycopene	singlet oxygen quencher
Ubiquinol-10	radical scavenger
Ascorbate (vitamin C)	diverse antioxidant functions
Glutathione (GSH)	diverse antioxidant functions
Urate	radical scavenger
Bilirubin	plasma antioxidant
Flavonoids	plant antioxidants (rutin, etc.)
Plasma proteins	metal binding, e.g. coeruloplasmin

TABLE 2   *(Continued)*

chemical	food additives, drugs
Enzymic (direct)	
superoxide dismutases	CuZn enzyme, Mn enzyme, Fe enzyme
glutathione peroxidases	enzymes (GPx, PHGPx)
	ebselen as enzyme mimic
catalase	heme protein, peroxisomes
Enzymatic (ancillary enzymes)	
conjugation enzymes	glutathione-S-transferases
	UDP-glucuronosyl-trans ferases
NADPH-quinone oxidoreductase	two-electron reduction
GSSG reductase	maintaining GSH levels
NADPH supply	NADPH for GSSG reductase
transport systems	GSSG export thioether (S-conjugate) export
repair systems	DNA repair systems oxidized protein turnover

6.3   HYALURONAN

In 1934, Karl Meyer and his colleague John Palmer isolated a previously unknown chemical substance from the vitreous body of cows' eyes. They found that the substance contained two sugar molecules, one of which was uronic acid. For convenience, therefore, they proposed the name "hyaluronic acid". The popular name is derived from "hyalos", which is the Greek word for glass + uronic acid [8]. At the time, they did not know that the substance which they had discovered would prove to be one of the most interesting and useful natural macromolecules. HA was first used commercially in 1942 when Endre Balazs applied for a patent to use it as a substitute for egg white in bakery products[9].

The term "hyaluronan" was introduced in 1986 to conform to the international nomenclature of polysaccharides and is attributed to Endre Balazs (Balazs E.A, et al., 1986), who coined it to encompass the different forms of the molecule can take, e.g., the acid form, hyaluronic acid, and the salts, such as sodium hyaluronate, which forms at physiological pH [10]. HA was subsequently isolated from many other sources and the physicochemical structure properties, and biological role of this polysaccharide were studied in numerous laboratories [11]. This work has been summarized in a Ciba Foundation Symposium [12] and a recent review [13].

Hyaluronan (HA) (Figure 2) is a unique biopolymer composed of repeating disaccharide units formed by N-acetyl-d-glucosamine and d-glucuronic acid. Both sugars are spatially related to glucose which in the beta configuration allows all of its bulky groups (the hydroxyls, the carboxylate moiety and the anomeric carbon on the adjacent sugar) to be in sterically favorable equatorial positions while all of the small hydrogen atoms occupy the less sterically favorable axial positions. Thus, the structure of the disaccharide is energetically very stable. HA is also unique in its size, reaching up to several million Daltons, and is synthesized at the plasma membrane rather than in the Golgi, where sulfated glycosaminoglycans are added to protein cores [14, 15].

In a physiological solution, the backbone of a HA molecule is stiffened by a combination of the chemical structure of the disaccharide, internal hydrogen bonds, and interactions with the solvent. The axial hydrogen atoms form a non-polar, relatively hydrophobic face while the equatorial side chains form a more polar, hydrophilic face, thereby creating a twisting ribbon structure. Solutions of hyaluronan manifest very unusual rheological properties and are exceedingly lubricious and very hydrophilic. In solution, the hyaluronan polymer chain takes on the form of an expanded, random coil. These chains entangle with each other at very low concentrations, which may contribute to the unusual rheological properties. At higher concentrations, solutions have an extremely high but shear-dependent viscosity. A 1% solution is like jelly, but when it is put under pressure it moves easily and can be administered through a small-bore needle. It has therefore been called a "pseudoplastic" material. The extraordinary rheological properties of hyaluronan solutions make them ideal as lubricants. There is evidence that hyaluronan separates most tissue surfaces that slide along each other. The extremely lubricious properties of hyaluronan, meanwhile, have been shown to reduce postoperative adhesion formation following abdominal and orthopedic surgery. As mentioned, the polymer in solution assumes a stiffened helical configuration, which can be attributed to hydrogen bonding between the hydroxyl groups along the chain. As a result, a coil structure is formed that traps approximately 1000 times its weight in water [16].

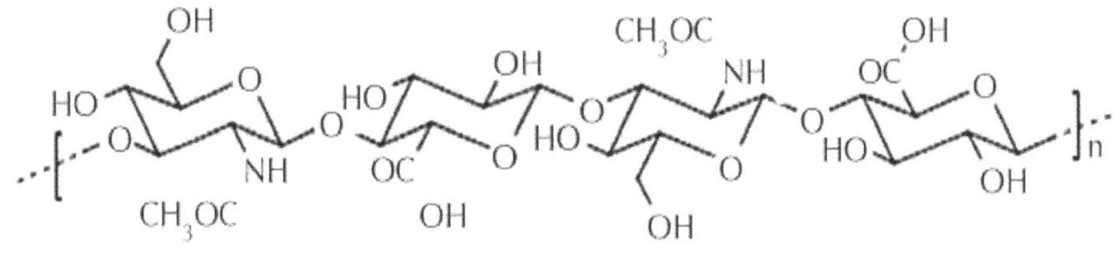

FIGURE 2   Structural formula of hyaluronan – the acid form.

6.3.1 HYALURONAN SOURCE

The HA is an important component of most connective tissues, including vitreous body, skin, synovial fluid, and umbilical cord. Comparative data for a variety of species, tissues and organs are shown in Table 3 [17]. It affects many cell functions, such as proliferation, differentiation and migration in a concentration and molar-mass dependent manner. The turnover of HA is extremely rapid. It is estimated that from 15 g of HA in the vertebrate body, 5 g turns over daily. The half-life of HA in the blood circulation is between 2-5 min. In the epidermis of the skin, where one half of HA of the body is found, it is up to 2 days, and in an apparently inert tissue as cartilage, it is approximate 1-3 weeks[18].

TABLE 3   Normal concentrations (µg/g) of hyaluronan (HA) in various organs of different species. (Laurent T C and Fraser J R E., 1996, Reed R K et al., 1988, Laurent UBC and Laurent T C., 1981, Laurent UBG., 1981)

Organ or fluid	Man	Sheep	Rabbit	Rat
Umbilical cord	4100			
Synovial fluid	1400-3600	540	3890	
Dermis	200			
Vitreous body	140-338	260	29	
Lung		98-243		34
Kidneys			93-113	30
Renal Papillae			250	
Renal cortex			4	
Brain	35-115		54-76	74
Muscles			27	
intestine				44
Thoracic lymph	8.5-18	1-34		5.4
Liver			1.5	4
Aqueous humour	0.3-2.2	1.6-5.4	0.6-2.5	0.2
Urine	0.1-0.3			
Lumbar CSF	0.02-0.32			
Plasma (serum)	0.01-0.1	0.12-0.31	0.019-0.086	0.048-0.26

The cellular synthesis of HA is a unique and highly controlled process. HA is naturally synthesized by a class of integral membrane proteins called hyaluronan synthases, of which vertebrates have three types: HAS1, HAS2, and HAS3 [19, 20]. Secondary structure predictions and homology modeling indicate an integral membrane protein (IMP). The IMP is a protein molecule (or assembly of proteins) that in most cases spans the biological membrane with which it is associated (especially the plasma membrane) or which, is sufficiently embedded in the membrane to remain with it during the initial steps of biochemical purification (in contrast to peripheral membrane proteins). Hyaluronan synthase enzymes synthesize large, linear polymers of the repeating disaccharide structure of hyaluronan by alternate addition of glucuronic acid and N-acetylglucosamine to the growing chain using their activated nucleotide sugars (UDP = glucuronic acid and UDP-N-acetyl glucosamine) as substrates.

6.3.1 HYALURONAN PRODUCTION

Currently there are two competing methods for industrial HA production that are extraction from animal sources, such as bovine eyes and rooster combs, and microbial production through the use of large scale fermentors. Both will be discussed in the following sections, in addition the opportunity of using novel genetically engineered microbial factories and a chemo-enzymatic synthesis approach will be reported from the very recent literature.

TRADITIONAL EXTRACTION PROCESSES

The traditional method for HA production is based on solvent extraction from animal tissue extracts, eventually using cetylpiridinium chloride (CPC) precipitation. One of the first paper presented by Swann (1968), reported the following procedure: (1) mechanical slicing of the rooster combs to obtain small pieces, (2) washing with ethanol (4 L ethanol to 1 Kg comb), this operation could be repeated until the solvent would not appear cloudy; (3) extracting the minced combs with a water/chloroform mixture (2.5 Kg combs: 10 L water: 0.5 L chloroform), while stirring to allow combs to swell; (4) filtering the solids from the broth and adding NaCl, successively carrying an additional chloroform extractions; (5) accomplishing protease (pronase) digestion, followed by chloroform extraction and centrifugation (Swann DA., 1968).

In alternative methods (29) the crude extracts were purified by epichlorohydrin triethanolamine- (ECTEOLA-) chromatography and by fractionation with CPC. In addition, repeated ethanol precipitation (1:3 water/ethanol ratio), before and after CPC (1%) HA precipitation, were reported [21]. In all the cases the product is then filtered through sterilizing filters, followed by solvent precipitation, finally HA is formulated into medical devices and pharmaceutical products. HA purified by these procedures was recovered with a yield greater than 90% with respect to the uronic acid evaluated in the starting material.

However, the collection of rooster combs and the extraction and purification procedures of HA from these tissues are time-consuming and labour intensive, making hyaluronan production very costly [22]. In fact in animal tissues hyaluronan is complexed with proteoglycans and often contaminated with HA degrading enzymes, making the isolation of high purity and high molecular sized polysaccharide very difficult. Moreover the use of animal-derived biomolecules for biopharmaceutical applications is facing growing opposition because of the risk of cross-species viral and other adventitious agent contaminations. Hence, since '80 microbial production is gradually replacing extraction from animal tissues in HA industrial manufacturing.

6.3.2 BIOTECHNOLOGICAL PRODUCTION OF HA

Bacteria known to be capable of the synthesis of HA are Streptococci of groups A and C, gram-positive bacteria such as *Streptococcus equi*, an equine pathogen, *Streptococcus equisimilis*, that is infective for different animals, *Streptococcus pyogenes*, a human pathogen and *Streptococcus uberis*, a bovine pathogen. These β-hemolytic bacteria, able to digest blood based agar medium, also present a slimy translucent layer surrounding bacterial colonies that can be attributed to HA synthesis. Figure (3) describes schematically steps of production of HA from Strepococcus species.

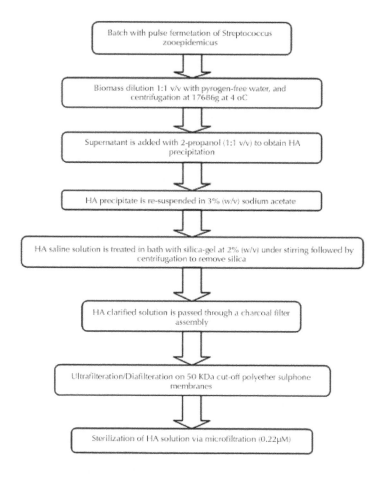

FIGURE 3 Overview of hyaluronic acid biotechnological production process from *Streptococcus zooepidemicus* fermentation to recently proposed downstream procedure as described by Rangaswamy and Jain (2008).

6.3.3   PROPERTIES OF HYALURONAN

HYALURONAN NETWORKS

The physico-chemical properties of hyaluronan were studied in detail from 1950 onwards (Comper WD and Laurent TC., 1978).

The molecules behave in solution as highly hydrated randomly kinked coils, which start to entangle at concentrations of less than 1 mg/mL. The entanglement point can be seen both by sedimentation analysis [23] and viscosity. (Morris ER et al., 1980) More recently Scott *et al. (1991)* have given evidence that the chains when entangling also interact with each other and form stretches of double helices so that the network becomes mechanically more firm.

RHEOLOGICAL PROPERTIES

Solutions of hyaluronan are viscoelastic and the viscosity is markedly shearing dependent. [24, 25] Above the entanglement point the viscosity increases rapidly and exponentially with concentration ($\sim c^{3.3}$) [26] and a solution of 10 g/L may have a viscosity at low shear of $\sim 10^6$ times the viscosity of the solvent. At high shear the viscosity may drop as much as $\sim 10^3$ times. [27] The elasticity of the system increases with increasing molecular weight and concentration of hyaluronan as expected for a molecular network. The rheological properties of hyaluronan have been connected with lubrication of joints and tissues and hyaluronan is commonly found in the body between surfaces that move along each other, for example, cartilage surfaces and muscle bundles. [28]

WATER HOMEOSTASIS

A fixed polysaccharide network offers a high resistance to bulk flow of solvent. [29, 30]This was demonstrated by Day, (1950) who showed that hyaluronidase treatment removes a strong hindrance to water flow through a fascia. Thus HA and other polysaccharides prevent excessive fluid fluxes through tissue compartments. Furthermore, the osmotic pressure of a hyaluronan solution is non-ideal and increases exponentially with the concentration. In spite of the high molecular weight of the polymer the osmotic pressure of a 10 g/L hyaluronan solution is of the same order as a 10g/L albumin solution. The exponential relationship makes hyaluronan and other polysaccharides excellent osmotic buffering substances — moderate changes in concentration lead to marked changes in osmotic pressure. Flow resistance together with osmotic buffering makes hyaluronan an ideal regulator of the water homeostasis in the body.

NETWORK INTERACTIONS WITH OTHER MACROMOLECULES

The hyaluronan network retards the diffusion of other molecules. [31]. It can be shown that it is the steric hindrance which restricts the movements and not the viscosity of the solution. The larger the molecule the more it will be hindered. In *vivo* hyaluronan will therefore act as a diffusion barrier and regulate the transport of other substances through the intercellular spaces. Furthermore, the network will exclude a certain volume of solvent for other molecules; the larger the molecule the less space will be available to it [32]. A solution of 10 g/L of hyaluronan will exclude about half of the solvent to serum albumin. Hyaluronan and other polysaccharides therefore take part in the partition of plasma proteins between the vascular and extravascular spaces. The excluded volume phenomenon will also affect the solubility of other macromolecules in the interstitium, change chemical equilibria and stabilize the structure of, for example, collagen fibers.

6.3.4   MEDICAL APPLICATIONS OF HYALURONIC ACID

The HA's viscoelastic matrix can act as a strong biocompatible support material and therefore is commonly used as growth scaffold in surgery, wound healing and embryology. In addition, administration of purified high molecular weight HA into orthopaedic joints can restore the desirable rheological properties and alleviate some of the symptoms of osteoarthritis. [33-36] The success of the medical applications of HA has led to the production of several successful commercial products, which have been extensively reviewed previously.

Table 4 summarizes both the medical applications and the commonly used commercial preparations containing HA used within this field. HA has also been extensively studied in ophthalmic, nasal and parenteral drug delivery. In addition, more novel applications including, pulmonary, implantation and gene delivery have also been suggested. Generally, HA is thought to act as either a mucoadhesive and retain the drug at its site of action/absorption or to modify the in vivo release/absorption rate of the therapeutic agent. A summary of the drug delivery applications of HA is shown in Table 5.

6.3.5   COSMETIC USES OF HYALURONIC ACID

The HA has been extensively utilized in cosmetic products because of its viscoelastic properties and excellent biocompatibility. Application of HA containing cosmetic products to the skin is reported to moisturize and restore elasticity thereby achieving an anti wrinkle effect, albeit no rigorous scientific proof exists to substantiate this claim. The HA-based cosmetic formulations or sunscreens may also be capable of protecting the skin against ultraviolet irradiation due to the free radical scavenging properties of HA. [37] HA, either in a stabilized form or in combination with other polymers, is used as a component of commercial dermal fillers (e.g. Hylaform®, Restylane® and Dermalive®) in cosmetic surgery. It is reported that injection of such products into the dermis, can reduce facial lines and wrinkles in the long term with fewer side-effects and better tolerability compared with the use of collagen. [38-42]. The main side-effect may be an allergic reaction, possibly due to impurities present in HA. [43]. Lin et al ((2000 have also investigated the feasibility of using HA as an alternative implant filler material to silicone gel in plastic surgery. These workers found that when using HA, the implanted organ structure was visually better than that obtained using silicone gel and saline implants, Moreover there were no reported in vivo side-effects 1 year after the implantation.

TABLE 4   Summary of the medical applications of hyaluronic acid (Brown MB and Jones SA., 2005)

Disease state	Applications	Commercial products	Publications
Osteoarthritis	Lubrication and mechanical support for the joints	Hyalgan® (Fidia, Italy), Artz® (Seikagaku, Japan) ORTHOVISC® (Anika, USA) Healon®, Opegan® and Opelead®	Hochburg, 2000, Altman, 2000, Dougados, 2000, Guidolin et al., 2001, Maheu et al., 2002, Barrett and Siviero, 2002, Miltner et al., 2002, Tascioglu and Oner, 2003, Uthman et al., 2003, Kelly et al., 2003, Hamburger et al., 2003, Kirwan, 2001, Ghosh and Guidolin, 2002, Mabuchi et al., 1999, Balazs, 2003, Fraser et al., 1993, Zhu and Granick, 2003.
Surgery and wound healing	Implantation of artificial intraocular lens Viscoelastic gel	Bionect®, Connettivina® and Jossalind®	Ghosh and Jassal, 2002, Risbert, 1997, Inoue and Katakami, 1993, Miyazaki et al., 1996, Stiebel-Kalish et al., 1998, Tani et al., 2002, Vazquez et al., 2003, Soldati et al., 1999, Ortonne, 1996, Cantor et al., 1998, Turino and Cantor, 2003.

TABLE 4   *(Continued)*

Embryo implantation	Culture media for the use of invitro fertilization	EmbryoGlue® (Vitrolife, USA)	Simon et al., 2003, Gardner et al., 1999, Vanos et al., 1991, Kemmann, 1998, Suchanek et al., 1994, Joly et al., 1992, Gardner, 2003, Lane et al., 2003, Figueiredo et al., 2002, Miyano et al., 1994, Kano et al., 1998, Abeydeera, 2002, Jaakma et al., 1997, Furnus et al., 1998, Jang et al., 2003.

TABLE 5   Summary of the drug delivery applications of hyaluronic acid

Route	Justification	Therapeutic agents	Publications
Ophthalmic	Increased ocular residence of drug, which can lead to increased bioavailability	Pilocarpine, tropicamide, timolol, gentimycin, tobramycin, arecaidine polyester, (S) aceclidine	Jarvinen et al., 1995, Sasaki et al., 1996, Gurny et al., 1987, Camber et al., 1987, Camber and Edman, 1989, Saettone et al., 1994, Saettone et al., 1991, Bucolo et al., 1998, Bucolo and Mangiafico, 1999, Herrero-Vanrell et al., 2000, Moreira et al., 1991, Bernatchez et al., 1993, Gandolfi et al., 1992, Langer et al., 1997.
Nasal	Bioadhesion resulting in increased bioavailability	Xylometazoline, vasopressin, gentamycin	Morimoto et al., 1991, Lim et al., 2002.
Pulmonary	Absorption enhancer and dissolution rate Modification	Insulin	Morimoto et al., 2001, Surendrakumar et al., 2003.
Parenteral	Drug carrier and facilitator of liposomal entrapment	Taxol, superoxide dismutase, human recombinant insulin-like growth factor, doxorubicin	Drobnik, 1991, Sakurai et al., 1997, Luo and Prestwich, 1999, Luo et al., 2000 Prisell et al., 1992, Yerushalmi et al., 1994, Yerushalmi and Margalit, 1998, Peer and Margalit, 2000, Eliaz and Szoka, 2001, Peer et al., 2003.

TABLE 5   *(Continued)*

Implant	Dissolution rate modification	Insulin	Surini et al., 2003, Takayama et al., 1990.
Gene	Dissolution rate modification and Protection	Plasmid DNA/mono- clonal antibodies	Yun et al., 2004, Kim et al., 2003.

6.3.6 BIOLOGICAL FUNCTION OF HYALURONAN

Naturally, hyaluronan has essential roles in body functions according to organ type that it distributes in it [44].

SPACE FILLER

The specific functions of hyaluronan in joints are still essentially unknown. The simplest explanation for its presence would be that a flow of hyaluronan through the joint is needed to keep the joint cavity open and thereby allow extended movements of the joint. Hyaluronan is constantly secreted into the joint and removed by the synovium. The total amount of hyaluronan in the joint cavity is determined by these two processes. The half-life of the polysaccharide at steady-state is in the order of 0.5–1 days in rabbit and sheep [44-46]. The volume of the cavity is determined by the pressure conditions (hydrostatic and osmotic) in the cavity and its surroundings. Hyaluronan could, by its osmotic contributions and its formation of flow barriers in the limiting layers, be a regulator of the pressure and flow rate. [47] It is interesting that in fetal development the formation of joint cavities is parallel with a local increase in hyaluronan. [48]

LUBRICATION

Hyaluronan has been regarded as an ideal lubricant in the joints due to its shear-dependent viscosity [49] but its role in lubrication has been refuted by others [50]. However, there are now reasons to believe that the function of hyaluronan is to form a film between the cartilage surfaces. The load on the joints may press out water and low-molecular solutes from the hyaluronan layer into the cartilage matrix. As a result the concentration of hyaluronan increases and a gel structure of micrometric thickness is formed which protects the cartilage surfaces from frictional damage [52]. This mechanism to form a protective layer is much less effective in arthritis when the synovial hyaluronan has both a lower concentration and a lower molecular weight than normal. Another change in the arthritic joint is the protein composition of the synovial fluid.

Fraser *et al* (1972) showed 25 years ago that addition of various serum proteins to hyaluronan substantially increased the viscosity and this has received a renewed interest in view of recently discovered hyaladherins (see above). The TSG-6 and inter-α-trypsin inhibitor and other acute phase reactants such as haptoglobin are concentrated to arthritic synovial fluid [52]. It is not known to what extent these are affecting the rheology and lubricating properties.

SCAVENGER FUNCTIONS

Hyaluronan has also been assigned scavenger functions in the joints. It has been known since the 1940s that hyaluronan is degraded by various oxidizing systems and ionizing irradiation and we know today that the common denominator is a chain cleavage induced by free radicals, essentially hydroxy radicals [53]. Through this reaction hyaluronan acts as a very efficient scavenger of free radicals. Whether this has any biological importance in protecting the joint against free radicals is unknown. The rapid turnover of hyaluronan in the joints has led to the suggestion that it also acts as a scavenger for cellular debris. [54] Cellular material could be caught in the hyaluronan network and removed at the same rate as the polysaccharide.

REGULATION OF CELLULAR ACTIVITIES

As discussed above, more recently proposed functions of hyaluronan are based on its specific interactions with hyaladherins. One interesting aspect is the fact that hyaluronan influences angiogenesis but the effect is different depending on its concentration and molecular weight [55]. High molecular weight and high concentrations of the polymer inhibit the formation of capillaries, while oligosaccharides can induce angiogenesis. There are also reports of hyaluronan receptors on vascular endothelial cells by which hyaluronan could act on the cells [56]. The avascularity of the joint cavity could be a result of hyaluronan inhibition of angiogenesis.

Another interaction of some interest in the joint is the binding of hyaluronan to cell surface proteins. Lymphocytes and other cells may find their way to joints through this interaction. Injection of high doses of hyaluronan intra-articularly could attract cells expressing these proteins. Cells can also change their expression of hyaluronan-binding proteins in states of disease whereby hyaluronan may influence immunological reactions and cellular traffic in the path physiological processes cells [57].The observation often reported that intra-articular injections of hyaluronan alleviates pain in joint disease [58] may indicate a direct or indirect interaction with pain receptors.

6.4   HYALURONAN AND SYNOVIAL FLUID

The synovial fluid, which consists of an ultrafiltrate of blood plasma and glycoproteins, in normal/healthy joint contains HA macromolecules of molar mass ranging between 6-10 mega Daltons [59]. SF serves also as a lubricating and shock absorbing boundary layer between moving parts of synovial joints. SF reduces friction and wear and tear of the synovial joint playing thus a vital role in the lubrication and protection of the joint tissues from damage during the motion [60].

As SF of healthy human exhibits no activity of the hyaluronidase, it has been inferred that oxygen-derived free radicals are involved in a self-perpetuating process of HA catabolism within the joint [61]. This radical-mediated process is considered to account for ca. twelve-hour half-life of native HA macromolecules in SF.

Acceleration degradation of high-molecular-weight HA occurring under inflammation and/or oxidative stress is accompanied by impairment and loss of its viscoelastic properties [62]. Low-molecular weight HA was found to exert different biological activities compared to the native high-molecular-weight biopolymer. The HA chains of 25–50 disaccharide units are inflammatory, immune-stimulatory, and highly angiogenic. HA fragments of this size appear to function as endogenous danger signals, reflecting tissues under stress [63-65]. Figure 4 describe the fragmentation mechanism of HA under free radical stress.

a) Initiation Phase: the Intact Hyaluronan Macromolecule Entering the Reaction with the HO˙ Radical Formed via the Fenton-Like Reaction:

$$Cu^+ + H_2O_2 \rightarrow Cu\ (II) + HO˙ + {}^-OH$$

H_2O_2 has its origin due to the oxidative action of the Weissberger_s system:

$$Asc + Cu^{2+} + 2\ O_2 + 4\ H^+ \rightarrow Asc + Cu^+ + 2\ H_2O_2.$$

b) Formation of an alkyl radical (C-centered hyaluronan macroradical) initiated by the HO˙ radical attack.

c) Propagation phase: formation of a peroxy-type C-macroradical of hyaluronan in a process of oxygenation after entrapping a molecule of O_2.

d) Formation of a hyaluronan-derived hydroperoxide via the reaction with another hyaluronan macromolecule.

e) Formation of Highly Unstable alkoxy- type C-macroradical of hyaluronan on Undergoing a Redox Reaction with a Transition Metal Ion in a Reduced State.

f) Termination phase: quick formation of alkoxy-type C-fragments and the Fragments with a terminal C=O group due to the glycosidic bond Scission of hyaluronan. Alkoxy-type C fragments may continue the propagation phase of the free-radical hyaluronan degradation reaction. Both fragments are represented by reduced molar masses [65-70].

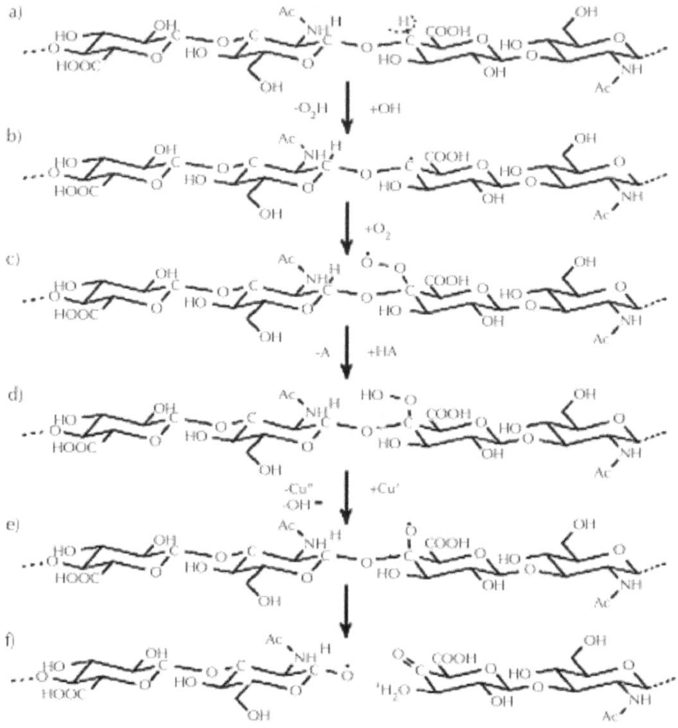

FIGURE 4   Schematic degradation of HA under free radical stress (E. Hrabarova et. al., 2012).

Several thiol compounds have attracted much attention from pharmacologists because of their reactivity toward endobiotics such as hydroxyl radical-derived species. Thiols play an important role as biological reductants (antioxidants) preserving the redox status of cells and protecting tissues against damages caused by the elevated reactive oxygen/nitrogen species (ROS/RNS) levels by which oxidative stress might be indicated.

Soltes and his coworkers examine the effect of several thiol compounds on inhibition of the degradation kinetics of a high-molecular-weight HA in vitro. High molecular weight hyaluronan (HA) samples were exposed to free-radical chain degradation reactions induced by ascorbate in the presence of Cu (II) ions the so called Weissberger's oxidative system. The concentrations of both reactants [ascorbate, Cu (II)] were comparable to those that may occur during an early stage of the acute phase of joint inflammation (see Figure 5). [75-80].

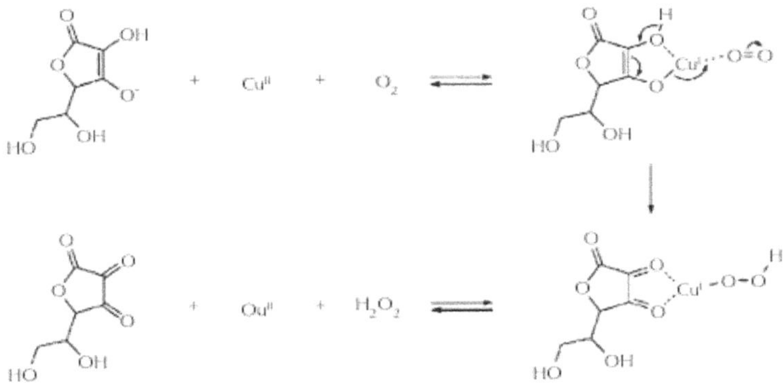

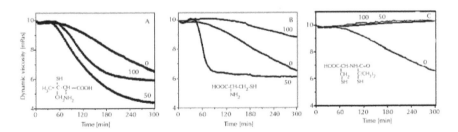

FIGURE 5   Scheme. Generation of H_2O_2 by Weissberger's System from Ascorbate and Cu (II) ions under Aerobic Conditions (Valachova K. et al., 2011).

Figure (6) illustrate the dynamic viscosity of hyaluronan solution in presence and absent of bucillamine, D-penicillamine and L-cysteine as inhibitors for free radical degradation of HA. The study shows that bucillamine has both a preventive and chain-breaking antioxidant. In the other hand D-penicillamine, and L-cysteine, dose dependently, acts as a scavenger of ˙OH radicals within the first 60 min then, however, its inhibition activity is lost and degradation of hyaluronan takes place[81-82]

FIGURE 6   Effect of A) L-Penicillamine, B) L-cysteine and C) Bucillamine with different concentration (50,100 μM) on HA degradation induced by the oxidative system containing 1.0 μM $CuCl_2$ + 100 μM ascorbic acid. (Valachova K et al., 2011).

L-Glutathione (GSH; L-γ-glutamyl-L-cysteinyl-glycine; a ubiquitous endogenous thiol, maintains the intracellular reduction oxidation (redox) balance and regulates signaling pathways during oxidative stress/conditions. GSH is mainly cytosolic in the concentration range of ca. 1–10 mM; however, in the plasma as well as in SF, the range is only 1–3 μM [83]. This unique thiol plays a crucial role in antioxidant defense, nutrient metabolism, and in regulation of pathways essential for the whole body homeostasis. Depletion of GSH results in an increased vulnerability of the cells to oxidative stress [84].

It was found that l-glutathione exhibited the most significant protective and chain-breaking antioxidative effect against the hyaluronan degradation. Thiol antioxidative activity, in general, can be influenced by many factors such as various molecule geometry, type of functional groups, radical attack accessibility, redox potential, thiol concentration and pK_a, pH, ionic strength of solution, as well as different ability to interact with transition metals. [85]

Figure (7) the dynamic viscosity versus time profiles of HA solution stressed to degradation with Weissberger's oxidative system. As evident, addition of different concentration of GSH resulted in a marked protection of the HA macromolecules against degradation. The greater the GSH concentration used, the longer was the observed stationary interval in the sample viscosity values. At the lowest GSH concentration used, i.e. 1.0 μM (Figure 7), the time-dependent course of the HA degradation was more rapid than that of the reference experiment with the zero thiol concentration. Thus, one could classify GSH traces as functioning as a pro-oxidant.

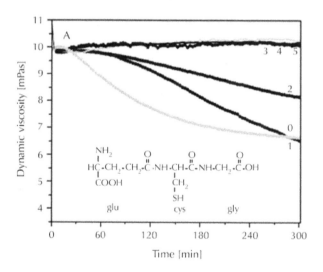

FIGURE 7   Comparison of the effect of L-glutathione on HA degradation induced by the system containing 1.0 mM $CuCl_2$ plus 100 μM L-ascorbic acid. Concentration of L-glutathione in μM: 1–1.0; 2–10; 3, 4, 5–50, 100, and 200, respectively. Concentration of Reference experiment: 0–nil thiol concentration (Hrabarova et al., 2009, Valachova et al., 2010a).

The effectiveness of antioxidant activity of 1, 4-dithioerythritol expressed as the radical scavenging capacity was studied by a rotational viscometry method [86]. L, 4-dithioerythritol widely accepted and used as an effective antioxidant in the field of enzyme and protein oxidation, is a new potential antioxidant standard exhibiting very good solubility in a variety of solvents. Figure (8) describe effect of 1, 4-dithioerythritol on degradation of HA solution under free radical stress [87].

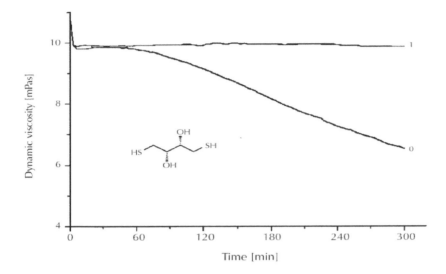

FIGURE 8   Effect of 1, 4-dithioerythritol on HA degradation induced by Weissberger's oxidative system containing (Hrabarova E et al 2010).

N-Acetyl-l-cysteine (NAC), another significant precursor of the GSH biosynthesis, has broadly been used as effective antioxidant in a form of nutritional supplement (Soloveva M. E et al 2007, Thibodeau P. A., et al 2001). At low concentrations, it is a powerful protector of α-1-antiproteinase against the enzyme inactivation by HOCl. NAC reacts with HO˙ radicals and slowly with H_2O_2; however, no reaction of this endobiotics with superoxide anion radical was detected [88].

An endogenous amine, cysteamine (CAM) is a cystine-depleting compound with antioxidative and anti-inflammatory properties; it is used for treatment of cystinosis – a metabolic disorder caused by deficiency of the lysosomal cystine carrier. CAM is widely distributed in organisms and considered to be a key regulator of essential metabolic pathways [89].

Investigation of the Antioxidative Effect of N-Acetyl-l-cysteine. Unlike l-glutathione, N-acetyl-l-cysteine was found to have preferential tendency to reduce Cu (II) ions to Cu (I), forming N-acetyl-l-cysteinyl radical (NAC.) that may subsequently react with molecular O_2 to give O_2^- (Soloveva M. E et al 2007, Thibodeau P. A., et al 2001). On the contrary to Cys, NAC (25 and 50 μM), when added at the beginning of the reaction, exhibited a clear antioxidative effect within ca. 60 and 80 min, respectively (Figure 9 (a)). Subsequently, NAC exerted a modest pro-oxidative effect, more profound at 25-μM than at 100-μM concentration (Figure 9 (a)). Application of NAC 1 h after the onset of the reaction (Figure 9 (b)) revealed its partial inhibitory effect against formation of the peroxy-type radicals, independently from the concentration applied [90].

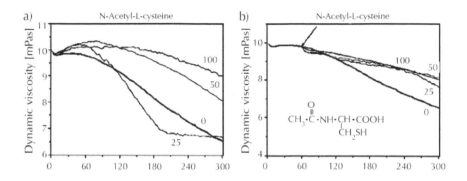

FIGURE 9 Evaluation of antioxidative effects of N-acetyl-l-cysteine against high-molar-mass hyaluronan degradation in vitro induced by Weissberger_s oxidative system. Reference sample (black): 1 µM Cu (II) ions plus 100 µM ascorbic acid; nil thiol concentration. N-Acetyl-l-cysteine addition at the onset of the reaction (a) and after 1 h (b) (25, 50,100 µM). (Hrabarova E, et al 2012).

Investigation of the Antioxidative Effect of Cysteamine. Cysteamine (100 µM), when added before the onset of the reaction, exhibited an antioxidative effect very similar to that of GSH (Figure 7 (a), and Figure 10 (a)). Moreover, the same may be concluded, when applied 1 h after the onset of the reaction (Figure 10 (b)), at the two concentrations (50 and 100 µM), suggesting that CAM may be an excellent scavenger of peroxy radicals generated during the per oxidative degradation of HA [91].

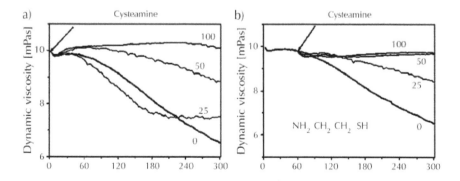

FIGURE 10 Evaluation of antioxidative effects of cysteamine against high-molar-mass hyaluronan degradation in vitro induced by Weissbergers oxidative system. Reference sample (black): 1 mM CuII ions plus 100 µM ascorbic acid; nil thiol concentration. Cysteamine addition at the onset of the reaction (a) and after 1 h (b) (25, 50,100 µM). (Hrabarova E, et al 2012).

ACKNOWLEDGEMENTS

The author would like to acknowledge Institute of Experimental Pharmacology & Toxicology, Slovak Academy of Sciences, at Bratislava, Slovakia for inviting and orienting him in the field of medical research. Also, he would like to thank Slovak Academic Information Agency (SAIA) for funding him during his work at the Institute.

KEYWORDS

- **Antioxidants**
- **Hyaluronan**
- **Oxidation stress**

REFERENCES

1. Abeydeera LR. In vitro production of embryos in swine. *Theriogenology* 2002; **57**: 257–273.
2. Adams ME. (ed.) Viseosupplementation: A treatment for osteoarthritis. *J. Rheumatol.* 1993; 20, Suppl. 39: 1-24.
3. Altman RD. Intra-articular sodium hyaluronate in osteoarthritis of the knee. *Semin Arthritis Rheum* 2000; **30**: 11–18.
4. Aruoma O. I., B. Halliwell, B. M. Hoey, J. Butler, Free Radic. Biol. Med.1989, 6, 593.
5. Ascorbate, and cupric ions. Neuroendocrinol. Lett. 29 (5), 2008a, 697-701.
6. Balazs E.A., Laurent T.C., Jeanloz R.W. (1986): Nomenclature of hyaluronic acid. Biochemical Journal, 235, 903.
7. Balazs EA, Denlinger JL. Clinical uses of hyaluronan. *Ciba Found Symp* 1989; **143**: 265–280.
8. Balazs EA, Denlinger JL. Viscosupplementation: a new concept in the treatment of osteoarthritis. *J Rheumatol* 1993; **20**: 3-9.
9. Balazs EA. Analgesic effect of elastoviscous hyaluronan solutions and the treatment of arthritic pain. *Cells Tissues Organs* 2003; **174**: 49–62.
10. Balazs, E. A.; Watson, D.; Duff, I. F.; Roseman, S. Arthritis Rheum. 1967, 10, 357–376.
11. Baňasová M., Valachová K., Hrabárová E., Priesolová E., Nagy M., Juránek I., Šoltés L.: Early stage of the acute phase of joint inflammation. In vitro testinf of bucillamine and its oxidized metabolite SA981 in the function of antioxidants. 16th Interdisciplinary Czech-Slovak Toxicological Conference in Prague, Interdisciplinary Toxicology, 4(2), 2011, p. 22.
12. Barrett JP, Siviero P. Retrospective study of outcomes in Hyalgan(R)-treated patients with osteoarthritis of the knee. *Clin Drug Invest* 2002; **22**: 87–97.
13. Bergeret-Galley C, Latouche X, Illouz YG. The value of a new filler material in corrective and cosmetic surgery: DermaLive and DermaDeep. *Aesthetic Plast Surg* 2001; **25**: 249–255.
14. Bernatchez SF, Tabatabay C, Gurny R. Sodium hyaluronate 0.25- percent used as a vehicle increases the bioavailability of topically administered gentamicin. *Graefes Arch Clin Exp Ophthalmol* 1993; **231**: 157–161.
15. Bothner H, Wik O. Rheology of hyaluronatc. Acta Otolaryngol. Suppl. (Sloekh.) 1987; 442: 25-30.

16. Brown MB, Jones SA.Hyaluronic acid: a unique topical vehicle for the localized delivery of drugs to the skin. JEADV (2005) 19 , 308–318

17. Brown TJ, Laurent UBG, Fraser JRE. Turnover of hyaluronan in synovial joints: elimination of labelled hyaluronan from the knee joints ofthe rabbit. *Exp. Physiol.* 1991; 76: 125-34.

18. Bucolo C, Mangiafico P. Pharmacological profile of a new topical pilocarpine formulation. *J Ocul Pharmacol Ther* 1999; **15**: 567– 573.

19. Bucolo C, Spadaro A, Mangiafico S. Pharmacological evaluation of a new timolol/pilocarpine formulation. *Ophthalmic Res* 1998; **30**: 101–106.

20. Camber O, Edman P, Gurny R. Influence of sodium hyaluronate on the meiotic effect of pilocarpine in rabbits. *Curr Eye Res* 1987; **6**: 779–784.

21. Camber O, Edman P. Sodium hyaluronate as an ophthalmic vehicle – some factors governing its effect on the ocular absorption of pilocarpine. *Curr Eye Res* 1989; **8**: 563–567.

22. Cantor JO, Cerreta JM, Armand G, Turino GM. Aerosolized hyaluronic acid decreases alveolar injury induced by human neutrophil elastase. *Proc Soc Exp Biol Med* 1998; **217**: 471–475.

23. Comper WD. Laurent TC. Physiological function of connective tissue polysaccharidcs. Physiol. Rev 1978; 58: 255-315.

24. Cowman M.K., Matsuoka S. (2005): Experimental approaches to hyaluronan structure. Carbohydrate Research, 340, 791–809.

25. Day TD. Connective tissue permeability and the mode of action of hyaluronidase. Nature 1950; 166: 785-6.

26. Dougados M. Sodium hyaluronate therapy in osteoarthritis: arguments for a potential beneficial structural effect. *Semin Arthritis Rheum* 2000; **30**: 19–25.

27. Dráfi F., Valachová K., Hrabárová E., Juránek I., Bauerová K., Šoltés L.: Study of methotrexate and β-alanyl-L-histidine in comparison with L-glutathione on high-molar-mass hyaluronan degradation induced by ascorbate plus Cu(II) ions via rotational viscometry. 60th Pharmacological Days in Hradec Králové, Acta Medica, 53 (3), 15.9.-17.9. 2010, p. 170.

28. Drobnik J. Hyaluronan in drug delivery. *Adv Drug Dev Rev* 1991; **7**: 295–308.

29. Duranti F, Salti G, Bovani B, Calandra M, Rosati ML. Injectable hyaluronic acid gel for soft tissue augmentation – a clinical and histological study. *Dermatol Surg* 1998; **24**: 1317–1325.

30. Edwards JCW et al. Consensus statement. Second international meeting on synovium. Cell biology, physiology and pathology. *Ann- Rheum. Dts.* 1995; **54**: 389-91.

31. Edwards JCW, Wilkinson LS, Jones HM et al. The formation of human synovial cavities: a possible role for hyaluronan and CD44 in altered interzone cohesion. J. *Anat.* 1994; **185:** 355-67.

32. Eliaz RE, Szoka FC. Liposome-encapsulated doxorubicin targeted to CD44: a strategy to kill CD44-overexpressing tumor cells. *Cancer Res* 2001; **61**: 2592–2601.

33. Figueiredo F, Jones GM, Thouas GA, Trounson AO. The effect of extracellular matrix molecules on mouse preimplantation embryo development in vitro. *Reprod Fertil Dev* 2002; **14**: 443–451.

34. Fisher A. E. Naughton O., D. P., Curr. Drug Delivery 2005, 2, 261.

35. Fraser J.R.E, Laurent T. C, Laurent U. B. G.Hyaluronan: its nature, distribution, functions and turnover.Journal of internal medicine 1997; 242: 27-33

36. Fraser JPE, Kimpton WG, Pierscionek BK, Cahill RNP. The kinetics of hyaluronan in normal and acute inflamed synovial joints: observations with experimental arthritis in sheep. Arthr Rheum 1993; 22 (suppl. 1):9-17.

37. Fraser JRE, Kimpton WG, Pierscionek BK, Cahill RNP. The kinetics of hyaluronan in normal and acutely inflamed synovial joints – observations with experimental arthritis in sheep. *Semin Arthritis Rheum* 1993; **22**: 9–17.
38. Fraser JRE, Laurent TC. Hyaluronan. In: comper WD, ed. Extracellular Matrix, 2. Molecular components and interactions. Amsterdam: Harwood academic Publications, 1996; pp. 141-99.
39. Fraser JRE. Foo WK. Maritz JS. Viscous interactions of hyaluronic acid with some proteins and neutral saccharides. *.Ann Rheutn. Dis.* 1972; 31: 513-20.
40. Fraser JRE. Kimpton WG, Pierseionek BK. Cahill RNP. The kineties of hyaluronan in normal and acutely inflamed synovial joints: observations with experimental arthritis in sheep. *Setnin. Arthritis Rheutn.* 1993; 22. Suppl. 1: 9-17.
41. Furnus CC, de Matos DG, Martinez AG. Effect of hyaluronic acid on development of in vitro produced bovine embryos. *Theriogenology* 1998; **49**: 1489–1499.
42. Gandolfi SA, Massari A, Orsoni JG. Low-molecular-weight sodium hyaluronate in the treatment of bacterial corneal ulcers. *Graefes Arch Clin Exp Ophthalmol* 1992; **230**: 20–23.
43. Gardner DK, Lane M, Stevens J, Schoolcraft WB. Changing the start temperature and cooling rate in a slow-freezing protocol increases human blastocyst viability. *Fertil Steril* 2003; **79**: 407–410.
44. Gardner DK, Rodriegez-Martinez H, Lane M. Fetal development after transfer is increased by replacing protein with the glycosaminoglycan hyaluronan for mouse embryo culture and transfer. *Hum Reprod* 1999; **14**: 2575–2580.
45. Ghosh P, Guidolin D. Potential mechanism of action of intraarticular hyaluronan therapy in osteoarthritis: are the effects molecular weight dependent? *Semin Arthritis Rheum* 2002; **32**: 10–37.
46. Ghosh S, Jassal M. Use of polysaccharide fibres for modem wound dressings. *Indian J Fibre Textile Res* 2002; **27**: 434–450.
47. Gibbs DA. Merrill EW, Smith KA, Balazs EA. Rheology of hyaluronic acid. Biopolymers 1968; 6: 777-91.
48. Glogau RG. The risk of progression to invasive disease. *J Am Acad Dermatol* 2000; **42**: S23–S24.
49. Grootveld M., E.B. Henderson, A. Farrell, D.R. Blake, H.G. Parkes, P. Haycock, Biochem. J.,273 (1991) 459-467.
50. Guidolin DD, Ronchetti IP, Lini E, *et al.* Morphological analysis of articular cartilage biopsies from a randomized. clinical study comparing the effects of 500–730 kDa sodium hyaluronate Hyalgan(R) and methylprednisolone acetate on primary osteoarthritis of the knee. *Osteoarthritis Cartilage* 2001; **9**: 371– 381.
51. Gurny R, Ibrahim H, Aebi A *et al.* Design and evaluation of controlled release systems for the eye. *J Control Release* 1987; **6**: 367–373.
52. Haddad J. J.,. Harb H. L, Mol. Immunol. 2005, 42, 987.
53. Halliwell B and Aruoma OI. (eds) DNA and Free Radicals, Boca Raton Press, 1993.
54. Halliwell, B. & Gutteridge, J. M. C. (1989) Free radicals in biology and medicine (2nd edn) Clarendon Press, Oxford.
55. Halliwell, B., Free Radicals, Antioxidants, and Human Disease: Curiosity, Cause, or Consequence? *Lancet* 1994; 344:721-724.
56. Hamburger MI, Lakhanpal S, Mooar PA, Oster D. Intra-articular hyaluronans: a review of product-specific safety profiles. *Semin Arthritis Rheum* 2003; **32**: 296–309.
57. Helmut SIES Strategies of antioxidant defense: review. Eur. J. Biochem. 215, 213-219 (1993)

58. Herrero-Vanrell R, Fernandez-Carballido A, Frutos G, Cadorniga R. Enhancement of the mydriatic response to tropicamide by bioadhesive polymers. *J Ocul Pharmacol Ther* 2000; **16**: 419–428.

59. Hlavacek M. The role of synovial fluid filtration by cartilage in lubrication of synovial joints. J. *Biomech.* 1993: 26: I 145-6U.

60. Hochberg MC. Role of intra-articular hyaluronic acid preparations in medical management of osteoarthritis of the knee. *Semin Arthritis Rheum* 2000; **30**: 2–10.

61. Hrabarova E , Katarina Valachova, Ivo Juranek, LadislavSoltes. Free-Radical Degradation of High-Molar-Mass Hyaluronan Induced by Ascorbate plus Cupric Ions: Evaluation of Antioxidative Effect of Cysteine- Derived Compounds, CHEMISTRY & BIODIVERSITY – Vol. 9 (2012).

62. Hrabarova E, Katar_na Valachova´, Peter Rapta, Ladislav Sˇ oltes. An Alternative Standard for Trolox-Equivalent Antioxidant-Capacity Estimation Based on Thiol Antioxidants. Comparative 2,2'-Azinobis[3- ethylbenzothiazoline-6-sulfonic Acid] Decolorization and Rotational Viscometry Study Regarding Hyaluronan Degradation. CHEMISTRY & BIODIVERSITY – Vol. 7 (2010).

63. Hrabarova E, Katarina Valachova, Jozef Rychly, Peter Rapta, Vlasta Sasinkova, Marta Malıkova, Ladislav Soltes. High-molar-mass hyaluronan degradation by Weissberger's system: Pro- and anti-oxidative effects of some thiol compounds. Polymer Degradation and Stability 94 (2009) 1867–1875.

64. Hrabarova E, Katarna Valachova, Ivo Juranek, LadislavSoltes. Free-Radical Degradation of High-Molar-Mass Hyaluronan Induced by Ascorbate plus Cupric Ions: Evaluation of Antioxidative Effect of Cysteine- Derived Compounds. CHEMISTRY & BIODIVERSITY – Vol. 9 (2012) 309-317

65. Hrabárová E., Valachová K., Juránek I., Šoltés L.: Free-radical degradation of high-molar-mass hyaluronan induced by ascorbate plus cupric ions. Anti-oxidative properties of the Piešťany-spa curative waters from healing peloid and maturation pool. In: "Kinetics, Catalysis and Mechanism of Chemical Reactions" G. E. Zaikov (eds), Nova Science Publishers, New York, 2011 pp. 29-36.

66. Hrabárová E., Valachová K., Rychlý J., Rapta P., Sasinková V., Gemeiner P., Šoltés L.: High-molar-mass hyaluronan degradation by the Weissberger´s system: pro- and antioxidative effects of some thiol compounds. Polym. Degrad. Stab. 94, 2009, 1867–1875.

67. Hultberg M., Hultberg B., Chem. Biol. Interact. 2006, 163, 192.

68. Hutadilok N. Ghosh P, Brooks PM. Binding of haptoglobin. inter-α-trypsin inhibitor, and l proteinase inhibitor to synovial fluid hyaluronate and the influence of these proteins on its degradation byoxygen derived free radicals. .*Ann Rheum Dis.* 1988; 47: 377-85.

69. Inoue M, Katakami C. The effect of hyaluronic-acid on corneal epithelial-cell proliferation. *Invest Ophthalmol Vis Sci* 1993; **34**: 2313–2315.

70. Itano N, Kimata K. Mammalian hyaluronan synthases. IUBMB Life 2002; 54: 195–199.

71. Jaakma U, Zhang BR, Larsson B, *et al.* Effects of sperm treatments on the in vitro development of bovine oocytes in semidefined and defined media. *Theriogenology* 1997; **48**: 711–720.

72. Jacob, R.A., The Integrated Antioxidant System. Nutr Res 1995; 15(5):755-766.

73. Jang G, Lee BC, Kang SK, Hwang WS. Effect of glycosaminoglycans on the preimplantation development of embryos derived from in vitro fertilization and somatic cell nuclear transfer. *Reprod Fertil Dev* 2003; **15**: 179–185.

74. Jarvinen K, Jarvinen T, Urtti A. Ocular absorption following topical delivery. *Adv Drug Dev Rev* 1995; **16**: 3–19.

75. Joly T, Nibart M, Thibier M. Hyaluronic-acid as a substitute for proteins in the deep-freezing of embryos from mice and sheep – an in vitro investigation. *Theriogenology* 1992; **37**: 473–480.

76. Kano K, Miyano T, Kato S. Effects of glycosaminoglycans on the development of in vitro matured and fertilized porcine oocytes to the blastocyst stage in vitro. *Biol Reprod* 1998; **58**: 1226–1232.

77. Kelly MA, Goldberg VM, Healy WL, *et al.* Osteoarthritis and beyond: a consensus on the past, present, and future of hyaluronans in orthopedics. *Orthopedics* 2003; **26**: 1064–1079.

78. Kemmann E. Creutzfeldt-Jakob disease (CJD) and assisted reproductive technology (ART) – quantification of risks as part of informed consent. *Hum Reprod* 1998; **13**: 1777.

79. Kessler A., M. Biasibetti, D. A. da Silva Melo, M. Wajner, C. S. Dutra-Filho, A. T. de SouzaWyse, C. M. D. Wannmacher, Neurochem. Res. 2008, 33, 737.

80. Kim A, Checkla DM, Dehazya P, Chen WL. Characterization of DNA-hyaluronan matrix for sustained gene transfer. *J Control Release* 2003; **90**: 81–95.

81. Kirwan J. Is there a place for intra-articular hyaluronate in osteoarthritis of the knee? *Knee* 2001; **8**: 93–101.

82. Kogan G., New Steps in Chemical and Biochemical Physics. Pure and Applied Science_, Eds. E. M. Pearce, G. Kirshenbaum, G. E. Zaikov, Nova Science Publishers, New York, 2011, p. 123.

83. Kreil G. (1995): Hyaluronidases-A group of neglected enzymes. Protein Sciences, 4, 1666–1669.

84. Lane M, Maybach JM, Hooper K, *et al.* Cryo-survival and development of bovine blastocysts are enhanced by culture with recombinant albumin and hyaluronan. *Mol Reprod Dev* 2003; **64**: 70–78.

85. Langer K, Mutschler E, Lambrecht G *et al.* Methylmethacrylate sulfopropylmethacrylate copolymer nanoparticles for drug delivery – Part III. Evaluation as drug delivery system for ophthalmic applications. *Int J Pharm* 1997; **158**: 219–231.

86. Langseth, L. From the Editor: Antioxidants and Diseases of the Brain. Antioxidant Vitamins Newsletter 1993;4:3.

87. Laurent C, Johnson-Wells G. Hellstrom S, Engstrom-Laurent A, Wells AF. Localization of hyaluronan in various muscular tissues. A morphological study m the rat. Cell Ti.ssue Res. 1991; 263: 201-5

88. LAURENT T C, ULLA BG LAURENT and J ROBERT E FRASER-' The structure and function of hyaluronan: An over view. Immunology and Cell Biology (1996) 74, A1-A7

89. Laurent T.C. (1989): The biology of hyaluronan. In: Ciba Foundation Symposium 143. John Wiley and Sons, New York. 1–298.

90. Laurent T.C., Fraser J.R.E. (1992): Hyaluronan. FASEB Journal, 6, 2397–2404.

91. Laurent TC, Fraser JRE. The properties and turnover of hyaluronan. In: functions of proteoglycans. Ciba foundation symposium 124. Chichester: wiley, 1986,9-29.

92. Laurent TC. Laurent UBG, Fraser JRE. Functions of hyaluronan. *Ann. Rheum. Dis.* 1995; **54**: 429-32.

93. Laurent TC. Ryan M. Pictruszkiewicz A. Fractionation of hyaluronic acid. The polydispersity of hyaluronic acid from the vitreous body. Biochim. Biophys. Acta 1960: 42: 476-85.

94. Laurent UBC, Laurent TC. On the origin of hyaluronate in blood.Biochem Int 1981;2:195-9.

95. Laurent UBG, Hyaluronate in aqueous humour. Exp Eye Res 1981;33: 147-55.

96. Lee J.Y., Spicer A.P (2000): Hyaluronan: a multifunctional, megaDalton, stealth molecule. Current Opinion in Cell Biology, 12, 581–586.

97. Leyden J et al Narins RS, Brandt F,. A randomized, double-blind, multicenter comparison of the efficacy and tolerability of Restylane versus Zyplast for the correction of nasolabial folds. *Dermatol Surg* 2003; **29**: 588–595.

98. Lim ST, Forbes B, Berry DJ, Martin GP, Brown MB. In vivo evaluation of novel hyaluronan/chitosan microparticulate delivery systems for the nasal delivery of gentamicin in rabbits. *Int J Pharm* 2002; **231**: 73–82.

99. Luo Y, Prestwich GD. Synthesis and selective cytotoxicity of a hyaluronic acid-antitumor bioconjugate. *Bioconjug Chem* 1999; **10**: 755–763.

100. Luo Y, Ziebell MR, Prestwich GD. A hyaluronic acid-taxol antitumor bioconjugate targeted to cancer cells. *Biomacromolecules* 2000; **1**: 208–218.

101. Maheu E, Ayral X, Dougados M. A hyaluronan preparation (500– 730 kDa) in the treatment of osteoarthritis: a review of clinical trials with Hyalgan(R). *Int J Clin Pract* 2002; **56**: 804–813.

102. Manuskiatti W, Maibach HI. Hyaluronic acid and skin: wound healing and aging. *Int J Dermatol* 1996; **35**: 539–544.

103. McDonald JN, Leviek JR. Effect of intra-articular hyaluronan on pressure-flow relation across synovium in anaesthetized rabbits. J. physiol. 1995; **485.1:** 179-93.

104. Meyer K., Palmer J.W. (1934): The polysaccharide of the vitreous humor. Journal of Biology and Chemistry, 107, 629–634.

105. Miltner O, Schneider U, Siebert CH, et al. Efficacy of intraarticular hyaluronic acid in patients with osteoarthritis – a prospective clinical trial. *Osteoarthritis Cartilage* 2002; **10**: 680–686.

106. Miyano T, Hirooka RE, Kano K et al. Effects of hyaluronic-acid on the development of 1-cell and 2-cell porcine embryos to the blastocyst stage in-vitro. *Theriogenology* 1994; **41**: 1299–1305.

107. Miyazaki T, Miyauchi S, Nakamura T, et al. The effect of sodium hyaluronate on the growth of rabbit cornea epithelial cells in vitro. *J Ocul Pharmacol Ther* 1996; **12**: 409–415.

108. Moreira CA, Armstrong DK, Jelliffe RW et al. Sodium hyaluronate as a carrier for intravitreal gentamicin – an experimental study. *Acta Ophthalmol (Copenh)* 1991; **69**: 45–49.

109. Moreira CA, Moreira AT, Armstrong DK et al. In vitro and in vivo studies with sodium hyaluronate as a carrier for intraocular gentamicin. *Acta Ophthalmol (Copenh)* 1991; **69**: 50–56.

110. Morimoto K, Metsugi K, Katsumata H, et al. Effects of lowviscosity sodium hyaluronate preparation on the pulmonary absorption of rh-insulin in rats. *Drug Dev Ind Pharm* 2001; **27**: 365–371.

111. Morimoto K, Yamaguchi H, Iwakura Y et al. Effects of viscous hyaluronate-sodium solutions on the nasal absorption of vasopressin and an analog. *Pharmacol Res* 1991; **8**: 471–474.

112. Morris ER. Rees DA, Welsh EJ. Conformation and dynamic interactions in hyaluronate solutions. J. Mol. Biol. 1980; 138: 383-400.

113. Myint P et al. The reaetivity of various free radicals with hyaluronie aeid; steady-state and pulse radioKsis studies. *Biochim. Biophys- Aeta* 1987; **925:** 194-202.

114. Necas J., Bartosikova L., Brauner P., Kolar J. Hyaluronic acid (hyaluronan): a review. Veterinarni Medicina, 53, 2008 (8): 397–411.

115. Noble P.W., Hyaluronan and its catabolic products in tissue injury and repair, Matrix Biol. 21 (2002) 25–29.

116. Oates K.M.N, W.E. Krause, R.H. Colby,Mat. Res. Soc. Syrnp. Proc.,711 (2002) 53-58.

117. Ogston AG, Stanier JE. The physiological function of hyaluronie acid in synovial fluid; viscous, elastic and lubrieant properties. J. *Physiol.* 1953; **199:** 244-52.

118. O'Regan M, Martini I, Crescenzi F, De Luca C, Lansing M, 1994. Molecular mechanisms and genetics of hyaluronan biosynthesis. Int J Biol Macromol 16(6): 283-6.

119. Ortonne JP. A controlled study of the activity of hyaluronic acid in the treatment of venous leg ulcers. *J Dermatol Treatment* 1996; 7: 75–81.

120. Parsons B.J., S. Al-Assaf, S. Navaratnam, G.O. Phillips, Comparison of the reactivity of different oxidative species (ROS) towards hyaluronan, in: J.F. Kennedy, G.O. Phillips, P.A. Williams, V.C. Hascall (Eds.), Hyaluronan: Chemical, Biochemical and Biological Aspects, Woodhead, Publishing Ltd., Cambridge, MA, 2002, pp. 141–150.

121. Peer D, Florentin A, Margalit R. Hyaluronan is a key component in cryoprotection and formulation of targeted unilamellar liposomes. *Biochim Biophys Acta-Biomembranes* 2003; **1612**: 76–82.

122. Peer D, Margalit R. Physicochemical evaluation of a stability-driven approach to drug entrapment in regular and in surface-modified liposomes. *Arch Biochem Biophys* 2000; **383**: 185–190.

123. Praest B.M., H. Greiling, R. Kock, Carbohydr. Res., 303 (1997) 153-151 .

124. Prescott AL, 2003. Method for purifying high molecular weight hyaluronic acid. USP 6660853.

125. Prisell PT, Camber O, Hiselius J, Norstedt G. Evaluation of hyaluronan as a vehicle for peptide growth factors. *Int J Pharm* 1992; **85**: 51–56.

126. Radin EL, Swann DA, Weisser PA. Separation of a hyaluronate-free lubricating fraction from synovial fluid. Nature 1970: **228**: 377-8.

127. Ramsarma T, Devasagayam T P A, Boloor K K. Methods for estimating lipid peroxidation: Analysis of merits and demerits (mini review). Indian J Bioche Biophys 2003; 40: 300-8.

128. Rangaswamy V and Jain D, 2008. An efficient process for production and purification of hyaluronic acid from streptococcus equi subsp. Zooepidemicus. Biotechnol Lett 30: 493-496.

129. Rapta P., Valachová K., Gemeiner P., Šoltés L.: High-molar-mass hyaluronan behavior during testing its antioxidant properties in organic and aqueous media: effects of the presence of Mn(II) ions. Chem. Biodivers. 6, 2009, 162-169.

130. Rapta P., Valachová K., Zalibera M., Šnirc V., Šoltés L.: Hyaluronan degradation by reactive oxygen species: scavenging eggect of the hexapyridoindole stobadine and two of its derivatives. In Monomers, Oligomers, Polymers, Composites, and Nanocomposites, , Ed: R. A. Pethrick; P. Petkov, A. Zlatarov; G. E. Zaikov, S. K. Rakovsky, Nova Science Publishers, N.Y., Chapter 7, 2010, pp. 113-126.

131. Reed R K, Lilja K, Laurent T C. Hyaluronan in the rat with special reference to the skin. Acta physiol scand 1988; 134:405-11.

132. Rees M. D., Kennett E. C., J. M. Whitelock, M. J. Davies, Free Radical Biol. Med. 2008, 44, 1973.

133. Risberg B. Adhesions: preventive strategies. *Eur J Surg* 1997; **163**: 32–39.

134. Rychly J.´, L. S olte´ s, M. Stankovska´ , I. Janigova´ , K. Csomorova´ , V. Sasinkova´ , G. Kogan, P. Gemeiner, Polym. Degrad. Stab. 2006, 91, 3174.

135. Saettone M F, Giannaccini B, Chetoni P, *et al.* Evaluation of highmolecular- weight and low-molecular-weight fractions of sodium hyaluronate and an ionic complex as adjuvants for topical ophthalmic vehicles containing pilocarpine. *Int J Pharm* 1991; **72**: 131–139.

136. Saettone M F, Monti D, Torracca M T, Chetoni P. Mucoadhesive ophthalmic vehicles – evaluation polymeric low-viscosity formulations. *J Ocul Pharmacol* 1994; **10**: 83–92.

137. Sakurai K, Miyazaki K, Kodera Y *et al.* Anti-inflammatory activity of superoxide dismutase conjugated with sodium hyaluronate. *Glycoconj J* 1997; **14**: 723–728.

138. Salasche SJ. Epidemiology of actinic keratoses and squamous cell carcinoma. *J Am Acad Dermatol* 2000; **42**: S4–S7.

139. Sasaki H, Yamamura K, Nishida K, et al. Delivery of drugs to the eye by topical application. Prog Retinal Eye Res 1996; 15: 583–620.

140. Sattar A, Kumar S, West DC. Does hyaluronan have a role in endothelial cell proliferation ofthe synovium. Semin. . Arthritis Rheum. 1992; 22: 37-43.

141. Schartz RA. The actinic keratoses. A perspective and update. Dermatol Surg 1997; 23: 1009–1019.

142. Scott JE, Cummings C, Brass A, Chen Y. Secondary and tertiary structures of hyaluronan in aqueous solution, investigated by rotary shadowing-electron microscopy and computer simulation. Biochem. J. 1991; 274: 600-705.

143. Sies, H. (1985) Oxidative stress, Academic Press, London.

144. Sies, H. et al., Antioxidant Function of Vitamins. Ann NY Acad Sci 1992; 669:7-20.

145. Simon A, Safran A, Revel A et al. Hyaluronic acid can successfully replace albumin as the sole macromolecule in a human embryo transfer medium. Fertil Steril 2003; 79: 1434–1438.

146. Soldati D, Rahm F, Pasche P. Mucosal wound healing after nasal surgery. A controlled clinical trial on the efficacy of hyaluronic acid containing cream. Drugs Exp Clin Res 1999; 25: 253–261.

147. Soloveva M. E, V. V. Solovev, A. A. Faskhutdinova, A. A. Kudryavtsev, V. S. Akatov, Cell Tissue Biol. 2007, 1, 40.

148. Šoltés L., Valachová K., Mendichi R., Kogan G., Arnhold J., Gemeiner P.: Solution properties of high-molar-mass hyaluronans: the biopolymer degradation by ascorbate. Carbohydr. Res. 342, 2007, 1071̄1077.

149. Stadtman ER. Protein oxidation and aging. Science 1992; 257: 1220-25.

150. Stiebel-Kalish H, Gaton DD, Weinberger D et al. A comparison of the effect of hyaluronic acid versus gentamicin on corneal epithelial healing. Eye 1998; 12: 829–833.

151. Suchanek E, Simunic V, Juretic D, Grizelj V. Follicular-fluid contents of hyaluronic-acid, follicle-stimulating-hormone and steroids relative to the success of in-vitro fertilization of human oocytes. Fertil Steril 1994; 62: 347–352.

152. Surendrakumar K, Martyn GP, Hodgers ECM, et al. Sustained release of insulin from sodium hyaluronate based dry powder formulations after pulmonary delivery to beagle dogs. J Control Release 2003; 91: 385–394.

153. Surini S, Akiyama H, Morishita M, et al. Polyion complex of chitosan and sodium hyaluronate as an implant device for insulin delivery. STP Pharm Sci 2003; 13: 265–268.

154. Surovcikova-Machova L., Valachova K., Banasova M., Snirc V., Priesolova E., Nagy M., Juranek I., Soltes L.: Free-radical degradation of high-molar-mass hyaluronan induced by ascorbate plus cupric ions: Testing of stobadine and its two derivatives in function as antioxidants. Gen. Physiol. Biophys., 2012, 31, 57–64.

155. Swann D A, 1968. Studies on hyaluronic acid: I. The preparation and properties of rooster comb hyaluronic acid. Bioch Bioph Acta (BBA) - General Subjects 156(1): 17-30.

156. Takayama K, Hirata M, Machida Y et al. Effect of interpolymer complex-formation on bioadhesive property and drug release phenomenon of compressed tablet consisting of chitosan and sodium hyaluronate. Chem Pharmaceut Bull 1990; 38: 1993–1997.

157. Tani E, Katakami C, Negi A. Effects of various eye drops on corneal wound healing after superficial keratectomy in rabbits. Jpn J Ophthalmol 2002; 46: 488–495.

158. Tascioglu F, Oner C. Efficacy of intra-articular sodium hyaluronate in the treatment of knee osteoarthritis. Clin Rheumatol 2003; 22: 112–117.

159. Thibodeau P. A., S. Kocsis-Be´dard, J. Courteau, T. Niyonsenga, B. Paquette, Free Radic. Biol. Med. 2001, 30, 62.
160. Trommer H, Wartewig S, Bottcher R et al. The effects of hyaluronan and its fragments on lipid models exposed to UV irradiation. Int J Pharm 2003; 254: 223–234.
161. Turino G M, Cantor J O. Hyaluronan in respiratory injury and repair. Am J Respir Crit Care Med 2003; 167: 1169–1175.
162. Uthman I, Raynauld J P, Haraoui B. Intra-articular therapy in osteoarthritis. Postgrad Med J 2003; 79: 449–453.
163. Valachova K, Eva Hrabarova, Elena Priesolova, Milan Nagy, Maria Ba˘nasova, Ivo Juranek, Ladislav Soltes. Free-radical degradation of high-molecular-weight hyaluronan induced by ascorbate plus cupric ions. Testing of bucillamine and its SA981-metabolite as antioxidants. J.Pharma & Biomedical Analysis 56 (2011) 664– 670.
164. Valachova K., Andrea Vargova, Peter Rapta, Eva Hrabarova´, Frantisek Drafi, Katarna Bauerova, Ivo Juranek, Ladislav Soltes. Aurothiomalate as Preventive and Chain-Breaking Antioxidant in Radical Degradation of High-Molar-Mass Hyaluronan. CHEMISTRY & BIODIVERSITY – Vol. 8 (2011) 1274-1283
165. Valachová K., Hrabárová E., Dráfi F., Juránek I., Bauerová K., Priesolová E., Nagy M., Šoltés L.: Ascorbate and Cu(II) induced oxidative degradation of high-molar-mass hyaluronan. Pro- and antioxidative effects of some thiols. Neuroendocrinol. Lett. 31 (2), 2010a, 101-104.
166. Valachová K., Hrabárová E., Gemeiner P., Šoltés L.: Study of pro- and anti-oxidative properties of d-penicillamine in a system comprising high-molar-mass hyaluronan,
167. Valachová K., Hrabárová E., Juránek I., Šoltés L. : Radical degradation of high-molar-mass hyaluronan induced by Weissberger oxidative system. Testing of thiol compounds in the function of antioxidants. 16th Interdisciplinary Slovak-Czech Toxicological Conference in Prague, Interdisciplinary Toxicology, 4(2), 2011b, p. 65.
168. Valachová K., Kogan G., Gemeiner P., Šoltés L.: Hyaluronan degradation by ascorbate: Protective effects of manganese(II). Cellulose Chem. Technol., 42 (9-10), 2008b, 473483.
169. Valachová K., Kogan G., Gemeiner P., Šoltés L.: Hyaluronan degradation by ascorbate: protective effects of manganese(II) chloride. In Progress in Chemistry and Biochemistry. Kinetics, Thermodynamics, Synthesis, Properties and Application, Nova Science Publishers, N.Y., Chapter 20, 2009b, pp. 201-215.
170. Valachová K., Mendichi R., Šoltés L.: Effect of l-glutathione on high-molar-mass hyaluronan degradation by oxidative system Cu(II) plus ascorbate. In Monomers, Oligomers, Polymers, Composites, and Nanocomposites, Ed: R. A. Pethrick; P. Petkov, A. Zlatarov; G. E. Zaikov, S. K. Rakovsky, Nova Science Publishers, N.Y., Chapter 6, 2010c, pp. 101-111.
171. Valachová K., Rapta P., Kogan G., Hrabárová E., Gemeiner P., Šoltés L.: Degradation of high-molar-mass hyaluronan by ascorbate plus cupric ions: effects of d-penicillamine addition. Chem. Biodivers. 6, 2009a, 389-395.
172. Valachová K., Rapta P., Slováková M., Priesolová E., Nagy M., Mislovičová D., Dráfi F., Bauerová K., Šoltés L.: Radical degradation of high-molar-mass hyaluronan induced by ascorbate plus cupric ions. Testing of arbutin in the function of antioxidant. In: "Advances in Kinetics and Mechanism of Chemical Reactions" G. E. Zaikov, A. J. M. Valente, A. L. Iordanskii (eds), Apple Academic Press, Waretown, NJ, USA 2013 pp. 1-19.
173. Valachová K., Šoltés L.: Effects of biogenic transition metal ions Zn(II) and Mn(II) on hyaluronan degradation by action of ascorbate plus Cu(II) ions. In New Steps in Chemical and Biochemical Physics. Pure and Applied Science, Nova Science Publishers, Ed: E. M.

Pearce, G. Kirshenbaum, G.E. Zaikov, Nova Science Publishers, N.Y., Chapter 10, 2010b, pp. 153-160.

174. Valachová K., Vargová A., Rapta P., Hrabárová E., Dráfi F., Bauerová K., Juránek I., Šoltés L.: Aurothiomalate in function of preventive and chain-breaking antioxidant at radical degradation of high-molar-mass hyaluronan. Chem. Biodivers., 8, 2011a, 1274-1283.

175. Vanos H C, Drogendijk A C, Fetter W P F, et al. The influence of contamination of culture-medium with hepatitis-B virus on the outcome of in vitro fertilization pregnancies. Am J Obstet Gynecol 1991; 165: 152–159.

176. Vazquez J R, Short B, Findlow A H et al. Outcomes of hyaluronan therapy in diabetic foot wounds. Diabetes Res Clin Pract 2003; 59: 123–127.

177. Weigel P H, Hascall V C, Tammi M. Hyaluronan synthases. J Biol Chem 1997; 272: 13997–14000.

178. West D.C., I.N. Hampson, F. Arnold, S. Kumar, Angiogenesis induced by degradation products of hyaluronic acid, Science 228 (1985) 1324–1326.

179. Yerushalmi N, Arad A, Margalit R. Molecular and cellular studies of hyaluronic acid-modified liposomes as bioadhesive carriers for topical drug-delivery in wound-healing. Arch Biochem Biophys 1994; 313: 267–273.

180. Yerushalmi N, Margalit R. Hyaluronic acid-modified bioadhesive liposomes as local drug depots: effects of cellular and fluid dynamics on liposome retention at target sites. Arch Biochem Biophys 1998; 349: 21–26.

181. Yoshikawa T, Toyokuni S, Yamamoto Y and Naito Y, (eds) Free Radicals in Chemistry Biology and Medicine, OICA International, London, 2000.

182. Yun Y H, Goetz DJ, Yellen P, Chen W. Hyaluronan microspheres for sustained gene delivery and site-specific targetting. Biomaterials 2004; 25: 147–157.

183. Zhu Y X, Granick S. Biolubrication: hyaluronic acid and the influence on its interfacial viscosity of an antiinflammatory drug. Macromolecules 2003; 36: 973–976.

CHAPTER 7

MAGNETIC NANOPARTICLES IN POLYMERS

R. A . DVORIKOVA, YU. V. KORSHAK, L. N. NIKITIN, M. I. BUZIN,
V. A. SHANDITSEV, Z. S. KLEMENKOVA, A. L. RUSANOV,
A. R. KHOKHLOV, A. LAPPAS, and A. KOSTOPOULOU

CONTENTS

7.1   INTRODUCTION

New magnetic nanomaterials have been synthesized from ferrocene-containing poly-phenylenes. The cyclotrimerization of 1,1□-diacetylferrocene by condensation reaction catalyzed by p-toluenesulfonic acid in the presence of triethyl orthoformiate both in solution and supercritical carbon dioxide in the temperature range of 70–200°C is described. The highly branched ferrocene-containing polyphenylenes prepared by this procedure were used as precursors for preparing magnetic nanomaterials. This was achieved by thermal treatment of polyphenylenes in the range of 200–750°C. The emerging of crystal magnetite nanoparticles of magnetite with the average size of 6 to 22 nm distributed in polyconjugated carbonized matrix was observed due to crosslinking and thermal degradation of polyphenylene prepolymers. Saturation magnetization of such materials came up to 32 $Gs \cdot cm^3/g$ in a filed of 2.5 kOe.

In the last few years, the enhanced interest to nanosized materials has been developed due to potentially unusual physical properties of those items compared to common substances. The magnetic properties of such substances are of great interest and presently under intense investigations [1-9].

The magnetically active materials are known to use for many purposes including data recording and information storage, for permanent magnets, in magnetic cooling systems, as magnetic sensors, and so on [1,2]. The precise drug delivery by using magnetically active nanoparticles is of current interest in biomedical field [10] as well as new probabilities may be created in the field of MR-imaging, cell sorting and cell separation, in bio-selection processes, for enzyme immobilizing, in immune analysis, in catalytic processes, and the like [11-14].

Stabilization of nanosized magnetic particles is a key problem and may be solved by their incorporation in oligomer or polymer matrixes–one of the possible ways applicable for preparation of new magnetic materials [1,2].

Previously, we have developed a new approach for nanosized composite preparation by thermal treatment of ferrocene-containing polymers with reactive terminal groups [15]. Thermal treatment of ferrocene-containing polymers prepared from 1,1¢-diacetylferrocene in the range of 150–350°C was accompanied by crosslinking and origination of iron-containing nanosized particles displaying magnetic-ordering properties. The synthesis of highly-branched ferrocene-containing polyphenylenes was performed also by cyclotrimerization of 1,1¢-diacetylferrocene in environmentally friendly solvent–liquid and supercritical carbon dioxide [16]. This medium had been much used as a "green environment" for conducting of various chemical processes including synthesis and modification reactions of polymers [17-24]. The new approaches for creation of metal-polymeric systems and magneto-active nanocomposites, in particular, are of special interest [25-33].

Using of p-toluenesulfonic acid (p-TSA) or $SiCl_4/C_2H_5OH$ mixture as catalysts for the reaction [16]. The reaction was performed in liquid and supercritical carbon dioxide in the presence of triethyl orthoformiate at pressure of 20 MPa and temperature of 20 and 50°C with the polymer yield around 20%. DTG-analysis revealed 5% weight loss of the polymer at 400°C with the weight of carbonized residue of 80% after heating at 750°C. When heated at 300°C, the process of crosslinking in the samples occurred and crystalline iron-containing nanosized magnetic particles with the average size of 10–40 nm with saturation magnetization of 13 Gs·cm^3/g in a magnetic field of 2.5 kOe were observed in a polymer matrix.

In the present chapter, the study of cyclotrimerization reaction of 1,1¢-diacetylferrocene has been carried out in solution and in supercritical carbon dioxide in a wide temperature range for obtaining higher polymer yields. The controlled thermal treatment of prepared ferrocene-containing polyphenylenes has been carried out for attaining improved magnetic properties.

7.2   EXPERIMENTAL

Ferrocene-containing polymers (FP) were prepared either by conventional solution method [15] or in supercritical carbon dioxide (SC-CO$_2$) (FPSC) using technique described in [16] at 20 MPa and temperatures varied from 70 to 200°C.

Conventional polymerization of the monomer was carried out in a flat-bottom flask supplied with a thermometer and magnetic stirrer. 1.0 g (3.7 mmol) of 1,1¢-diacetylferrocene, 3 ml (8.0 mmol) of triethyl orthoformiate, and 0.10 g p-toluenesulfonic acid were placed into the flask and allowed to react at 70°C during 2.5 hrs and then left for 40 hrs at room temperature. The dark brown precipitate was filtered, washed with water to neutral reaction, then with ethanol and dried in vacuum. The yield was 0.34 g (24% from theory). Increasing of the temperature up to 140°C raised yields to 71%. Polymerization in SC-CO$_2$ was performed in a high pressure reactor with the inner capacity of 10 ml. The reactor was flushed out with CO$_2$ after loading of the reagents and heated up to the required temperature (±0.5°C). The input of CO$_2$ and applying of the required pressure (20 MPa) was completed by hand-operating press ("High Pressure Equipment", USA). The reaction mixture was agitated with magnetic stirrer and the reaction was carried out from 2.5 to 5 hr. After completing of the reaction the autoclave was cooled down and the pressure released. The resulting polymer was washed successively with ethanol and water to pH = 6–7 and then dried. The maximum yield was almost quantitative.

Magnetoactive materials were prepared by heating of the original samples of FP and FPSC in quartz tubes placed into measurement cell of magnetometer in the range of 200–750°C, or by thermal treatment of the samples in glass tubes under argon in the range of 250–500°C.

Appearance of the magnetic order during thermal treatment of the polymers was monitored by vibration magnetometer of the Foner type.

Morphology of the nanocomposites was studied by TEM using LEO 912AB OMEGA instrument. Size distribution of nanoparticles was figured out by statistical treatment of 50–100 particle size.

X-ray powder diffraction was obtained with Rigaku D/MAX-2000H rotating anode diffractometer with CuKα radiation and secondary pyrolytic graphite monochromator. DC magnetic susceptibility measurements were carried out with an Oxford Instruments MagLab EXA extraction magnetometer.

TG analysis was carried out with "DERIVATOGRAPH-C" (MOM, Hungary) instrument in air and argon by using samples of ~ 15 mg at a heating rate 5°C/min.

7.3   RESULTS AND DISCUSSION

The cyclotrimerization reaction of 1,1¢-diacetylferrocene catalyzed by p-TSA was studied for determining the optimal conditions for higher yields of polyphenylenes as precursors for materials with improved magnetic properties.

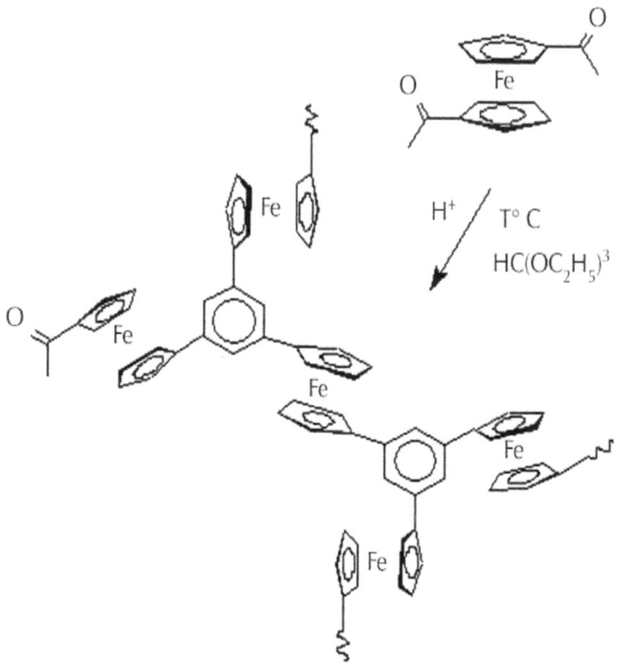

SCHEME 1   Reaction procedure.

Ferrocene-containing polymers were synthesized both under conventional conditions and in supercritical carbon dioxide (SC–CO$_2$), p-TSA was used as a catalyst for cyclotrimerization of 1,1¢-diacetylferrocene. The reaction proceeded in accordance with the Scheme 1 [16] which did not consider side reactions that could bring about some defect units of dypnone (β-methylchalcone) structures [34].

The effect of reaction conditions, such as temperature, duration, catalyst concentration, and post-stirring period at ambient temperature on yield and properties of the final products was studied. Tables 1 and 2 present some data on the reaction conditions and some properties of the polyphenylenes prepared under conventional conditions (FP) and in supercritical carbon dioxide (FPSC). The reaction was carried out in the temperature range of 70-200°C in the presence of triethyl orthoformate, which acts simultaneously as a ketalizing agent and as a solvent. Besides, ditolyl methane was used in some experiments as a solvent. The resulting ferrocene-containing polyphenylenes were obtained as powders of dark-brown color partially soluble in organic solvents such as dioxane, methylene chloride and benzene.

The yields of polymers increased at 100°C from 24 to 66%. Elemental analysis showed good agreement between found and calculated data and, in addition, sulfur was found in the samples in ~2% amount. The presence of the latter may be due to the chemical interaction between polymer and p-NSA that may also explain the reduced content of iron and carbon in analysis data. The maximum yield of FP was achieved at 100°C and 10% concentration of a catalyst, whereas 20% catalyst concentration provided the highest yields at 140°C.

TABLE 1 Synthesis of ferrocene-containing polymers (FP) with p-TSA as a catalyst and specific magnetization values of the samples

| Sample | Temperature, °C | Duration/Postreaction time at ambient temperature, hrs | Catalyst, % | Yield, % | Elemental analysis, %* | | | | $\sigma_H max$, G·cm³/g (at temperature) |
					C	H	Fe	S	
FP-1	70	2.5/40	10	24	66.51	4.86	20.69	0.80	9(500°C)
FP-2	70	5/20	10	25	65.64	4.91	20.02	0.62	10(500°C)
FP-3	100	2	10	16	64.43	4.56	12.83	2.11	21(750°C)
FP-4	100	2	20	24	61.51	4.34	13.10	3.18	21(500°C)
FP-5	100	2/15	10	45	68.81	5.45	18.55	-	12(500°C)
FP-6	100	2/20	10	66	62.85	5.05	19.37	0.65	26(500°C)
Sample	Temperature, °C	Duration/Postreaction time at ambient temperature, hrs	Catalyst, %	Yield, %	Elemental analysis, %*				$\sigma_H max$, G·cm³/g (at temperature)
					C	H	Fe	S	
FP-8	120	2	10	19	64.95	4.73	12.81	2.11	18(1000°C)
FP-9**	130	2/20	10	19	50.14	4.15	26.67	2.00	21(750°C)
FP-10	140	2/20	20	71	67.32	5.51	20.2	-	32(750°C)

*Calculated for $C_{126}H_{102}Fe_9O_6$: C – 66.88%, H – 4.54%, Fe - 22.21%
**Reaction was carried out in ditolylmethane; Molar ratio TEOF/Acetyl group was 1.2

The increase of the reaction time at 100°C resulted in an enhancing yield from 16 to 66% (Table 1). The same effect was observed when the catalyst concentration was raised up from 10 to 20%.

TABLE 2   Preparation of FPSC polymers in supercritical CO_2 and specific magnetization values of the samples after heat treatment at 500°C

Sample	Temperature, °C	Duration/ Postreaction time at ambient temperature, Hrs	Catalyst, %	Yield, %	Elemental analysis, %				Magnetization $(\sigma_{H^{o}max})$ G·cm³/g,
					C	H	Fe	S	
FPSC-1	70	2.5/20	10	32	67.43	5.17	20.76	0.25	6.0
FPSC -2	140	2/20	10	31	58.98	4.40	16.47	2.73	15.0
FPSC -3	140	2/20	60	98	53.67	4.68	23,36	0	12.5
FPSC -4	160	2/20	10	37	62.09	4.44	16.04	2.01	12.5
FPSC -5	200	2	10	92	63.43	4.95	16.81	0.80	14.5

Preparation of ferrocene-containing polymers in SC-CO_2 medium (FPSC) was of special interest. Some data are given in Table 2. Thus, the increase in reaction temperature from 70 to 200°C rose up the polymer yield from 10 to 90%, whereas the increasing the catalyst concentration from 10 to 60% brought about the quantitative yield of polyphenylenes.

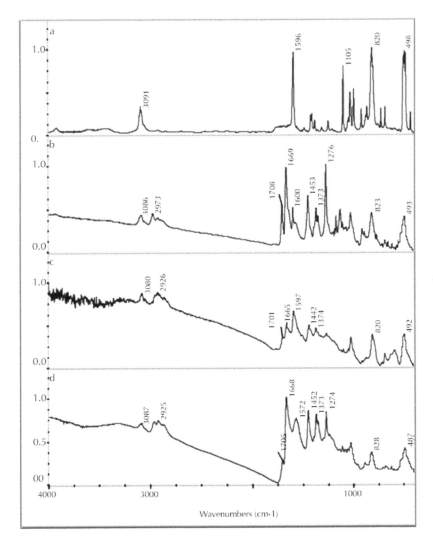

FIGURE 1   The IR-spectra of: (a) 1,3,5-triferrocenylbenzene; (b) FP-10 (Table 1), (c) FP-10 after heating at 350°C for 1 hr in measurement cell of magnetometer; (d) FPSC-5 (Table 2).

IR-spectra of FP revealed the band at 1600 $см^{-1}$ typical for stretching modes of CH-groups in 1,3,5-substituted benzene rings (Figure 1 (b)), while this band was observed at 1572 $см^{-1}$ for FPSC (Figure 1 (d)). The bands typical for ferrocenyl units were also found IR-spectra of the polymers (stretching modes of CH-groups at 3090 $см^{-1}$ non-planar bending vibrations of CH-groups in C_p rings at 820-830 $см^{-1}$, and the bands for doubly degenerated antisymmetric stretching modes Fe-C_p at 484 $см^{-1}$). Strong bands at 1275 $см^{-1}$ are due to symmetric stretching modes of C–C in disubstituted C_p rings and intensive bands at 1670 $см^{-1}$ и 1700 $см^{-1}$ may be referred to C=O stretching vibrations in dypnone fragments and terminal acetyl groups. IR-spectrum of model 1,3,5-triferrocenyl benzene shows the same characteristic bands typical for the polymer samples (Figure 1 (a)) except visible broadening of the latter due to higher molecular weight.

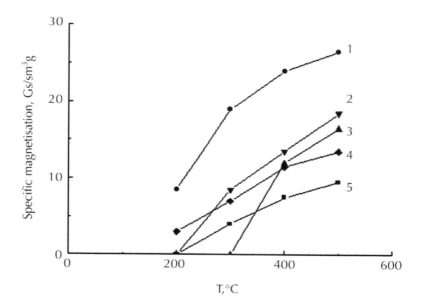

FIGURE 2   Magnetization curves for FPs obtained at different temperature: 1) FP-6; 2) FP-9; 3) FP-7; 4) FP-8; 5) FP-1.

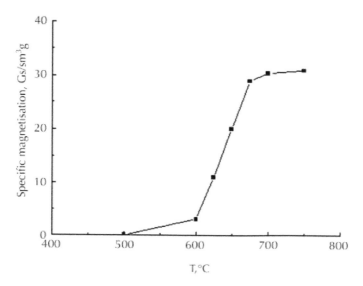

FIGURE 3   Magnetization curve for FP-10.

The amount of the β-methylchalcone fragments and acetyl end groups in polymers was found to reduce after heating samples at 250°C and higher temperatures as evidenced by decreasing intensities of the bands at 1600 cм⁻¹ and 1700 cм⁻¹. Simultaneous increase of the regular units in polymer molecules was found due to presence of 1,3,5-substituted benzene rings as evidenced by increasing of the band at 1600 cм⁻¹ specific for those groups (Figure 1 (c)). At the same time, network formation occurred and arising of the crystal iron-containing particles was observed. The mechanism of the particle formation comprises their spontaneous adjustment at the nanosized level depending on the polymer structure. Magnetic behavior of the final products is markedly influenced both by temperature regimes during ferrocene-containing polyphenylenes synthesis and subsequent thermal treatment. Thus, formation of magnetic phase in sample FP-6 (Figure (2)) prepared by conventional method at 100°C (FP) begins after 200°C and the maximum magnetization is equal to 26 Gs·cm³/g. Meanwhile, the sample FP-10 that was synthesized at 140°C showed the arising of magnetic phase after heating at 500°C, and at 750°C the magnetization approached to its maximum value of 32 Gs·cm³/g (Figure 3).

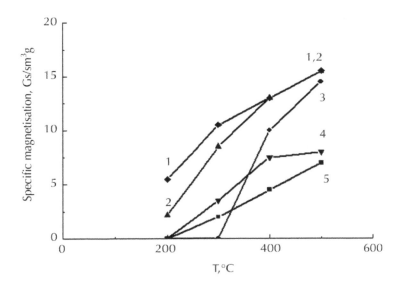

FIGURE 4   Magnetization curves for FPSCs obtained at different temperatures: 1) FPSC-2; 2) FPSC-4; 3) FPSC-5; 4) FPSC-3; 5) FPSC-1.

Figure 4 presents the plot of magnetization values versus heat treatment temperature for ferrocene-containing polyphenylenes synthesized in the range of 70–200°C in supercritical carbon dioxide (FPSC). For such samples formation of magnetic phase stats at 200°C for samples FPSC-2 and FPSC-4 prepared at 140 and 160°C, respectively. The highest magnetization (~15 Gs·cm³/g) was achieved for samples FPSC-2, FPSC-4 and FPSC-5, which were synthesized at 140,160, and 200°C, respectively (Table 2). However, FPSC-5 shows magnetic behavior just at 300°C. The lowest magnetization value after heat treatment was recorded for sample FPSC-1 originally prepared at 70°C. Besides, the samples synthesized in SC-CO$_2$ medium had the reduced magnetic properties compared to FP that could be explained by their higher thermostability.

FIGURE 5   Microphotographs of: (a) FPSC-5 after heating in argon at 250°C for 2 hr, average size of nanoparticles – 8 nm; (b) FP-8 after heating in argon at 350°C for 2 hr, average size of nanoparticles – 13 nm, and (c) FP-6 after heating at 500°C for 1 hr (heated in measurement cell of magnetometer), average size of nanoparticles – 22 nm.

TEM study of thermally treated ferrocene-containing polyphenylenes revealed in all samples presence of iron nanoparticles homogeneously distributed in polymer matrix with the average size of 6 to 22 nm depending on the synthesis conditions and heating regime. It should be mentioned that by increasing temperature of thermal treatment from 250 to 500°C one could increase the size of nanoparticles from 8 to 22 nm for the samples obtained in the presence of p-TSA as a catalyst. TEM images are shown in Figure 5 (a) and Figure 5 (b) for samples FPSC-5 and FP-8 subjected to heating at 250 and 350°C, respectively, in argon for 2 hr. The size of nanoparticles increased when the temperature of heat treatment rose up. Figure 5 (c) shows TEM image revealing nanoparticles with the average size of 22 nm for FP-6 sample after heating at 500°C in argon for 1 hr.

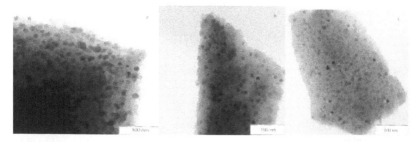

FIGURE 6   Microphotographs for FP-6 (a) and FP-10 (b, c), heated in measurement cell of magnetometer for 1 hr at 500 (a, b) and 700°C (c). The average size of particles – 22 (a), 10.63 (b), and 6.32 nm (c).

In the meantime, the particle size in a sample prepared with 20% of a catalyst and treated at the same temperature of 500°C was half as much and equal to 10,6 nm (Figure 6 (a)) and the subsequent heating up at 700°C resulted in reducing to 6.3 nm (Table 3, Figure 6 (b)).

TABLE 3 Size distribution of magnetic nanoparticles as a function of heat treatment temperature

Heat trearment temperature, °C	500	600	650	675	700
Average nanoparticle size, nm	10.63	9.21	6.96	6.66	6.32
Maximum size of nanoparticles, nm	13.63	15.21	25.23	22.07	20.56
Minimum size of nanoparticles, nm	8.55	4.17	2.35	2.09	2.14

Table 3 presents the data on nanoparticle size distribution for FP-7 sample after heat treatment at 500, 600, 650, 675, and 700°C. It is evident that increasing of temperature brought about the visible effect on the decreasing of the mean size of nanoparticles. Moreover, the particle size distribution for the sample was rather narrow when it was kept at 500°C compared to higher temperatures.

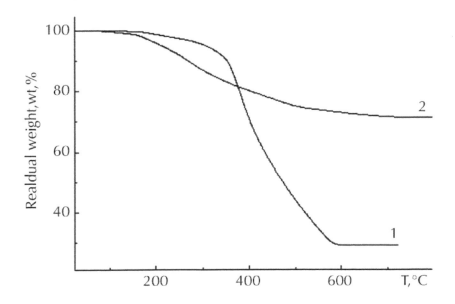

FIGURE 7 TGA curves for FP-10 in air (1) and in argon (2) at a heating rate 5°C/min.

Thermal and thermooxidative stability of synthesized ferrocene-containing polymers was studied by TGA method. It was found that thermal and thermooxidative degradation followed the basic features of such processes. Thus, the higher degree of decomposition in air was observed for all investigated polymers compared to that in

an inert atmosphere. As an example, Figure 7 shows TGA curves for FP-10 sample. The weight loss of the sample in air was about 70% whereas in inert atmosphere the sample loses essentially less from the initial weight (about 30%). The weight loss, both in air and in argon was practically coinciding and begins at the same temperature (190°C). Then, the thermal decomposition under heating in air was slowed down, and the sample starts to lose weight intensively only in the vicinity of 400°C. Such a behavior can be attributed to a formation of the dense network in the polymeric matrix at the initial stage of the thermooxidative degradation that prevents thermal degradation for a while. However, the peroxides formed during the polymer oxidation in air may form a redox systems together with ferrous species in ferrocene units to yield additional amounts of free radicals that finally stimulate a deeper degradation of the polymer. On the contrary, in an inert atmosphere the sample slowly loses weight in a wide temperature interval, up to 700°C and then its weight does not change.

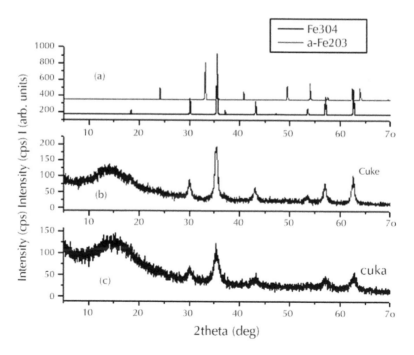

FIGURE 8 X-ray powder diffraction (XRD) patterns of two ferrocene-containing polyphenylene samples FP-6 (b), and FPSC-5 (c) after heating at 250°C, and the model spectra of two iron-based oxides (a).

The most significant observation concerns the comparison of the XRD spectra for the samples prepared by conventional way and in SC-CO$_2$ (FP-6 and FPSK-5) and heated at 250°C, and the models shown in Figure 8 (c). We may conclude from Figure 8(a)–(c) that the only crystalline phase presents in the samples if that of magnetite Fe$_3$O$_4$ and there is an excellent match between the experimental data and Fe$_3$O$_4$ model.

In addition to this crystalline phase there is a significant volume fraction arising from an amorphous phase, which is identified by the diffuse scattering (broad hump) in the 2Θ range 10-20 degrees. Based on the excess broadening of the Bragg reflections of Fe$_3$O$_4$, in sample FPSC-5, we suggest that the magnetite is a little less crystalline than that present in FP-6. Probably this is due to the different conditions of the synthesis for this material, since the sample was prepared in SC-CO$_2$ medium.

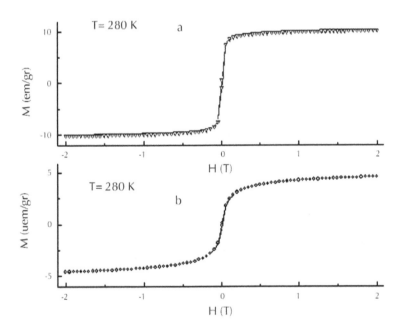

FIGURE 9   Magnetization curves for FP-6 (a) and FPSC-5 (b) at room temperature.

To this extent the following results, on the magnetic properties of these samples, are mainly dictated by the behavior of the magnetite whichis magnetic (ferromagnetic) already at room temperature (Figure 9). Although the measured magnetization, at T~ 280 K, is characteristic of a soft ferromagnetic-like material, with little coercivity (H$_c$ a few Gauss), there are a some qualitative differences (e.g. steep change of M vs. H for FP-6) between the curves of the two materials, probably due to the higher sample crystallinity (FP-6) or/and larger particle size.

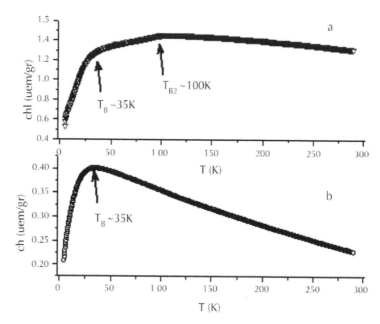

FIGURE 10 Temperature dependence of the zero-field cooled (ZFC) magnetic susceptibility of FP-6 (a) and FPSK-5 (b). The arrows indicate various "blocking" temperatures (TB) in these samples.

As far as the magnetic susceptibility is concerned (Figure 10), we find an important reduction of the magnetic moment at relatively low temperature ($T<T_B$) for both samples. Furthermore, FP-6 displays a second, higher "blocking" temperature ($T_{B2}\sim$ 100 K) which may be related to the Verwey transition (~118 K) met in bulk magnetite–shifted to relatively lower-T here maybe due to smaller particle-size.

CONCLUSION

The optimal conditions for synthesis of highly branched ferrocene-containing polyphenylenes were developed via cyclotrimerization of 1,1¢-diacetylferrocene by condensation reaction in the range of 70–140°C providing 71% yield of the polymer. The quantitative yields of the polymers were obtained in supercritical carbon dioxide (SC-CO$_2$).

Magnetic nanocomposites with saturation magnetization of 32 Gs·cm³/g in a field of 2.5 kOe were prepared by thermal treatment of such polymers in the range of 250–750°C. TGA data approved that initial temperature of sample degradation fit in with magnetic phase formation. XRD studies and magnetic measurements showed that crystalline phase of magnetite Fe$_3$O$_4$ presents in the samples with magnetization characteristic of a soft ferromagnetic-like material with small coercivity.

TEM study of polymer samples approved that the mean size of magnetic nanoparticles in polymer matrixes may be controlled by varying reaction conditions and heat treatment regime of the final polymers.

REFERENCES

1. Pomogaylo, A. D., Rosenberg, A. S., and Uflyand, A. S. *Metal Nanoparticals in Polymers. M.*, Khimiya, p 672 (2000).
2. Gubin, S. P., Koksharov, Yu. A., Khomutov, G. B., Yurkov, G. Yu. *Russ. Chem. Revs.*, **74**(6), 489–520 (2005).
3. Gudoshnikov, S., Liubimov, B., Matveets, L., Ranchinski, M., Usov, N., Gubin, S., Yurkov, G., Snigirev, O., and Volkov, I. *Journal of Magnetism and Magnetic Materials*, **258–259**, 54–56(2003).
4. Baker, C., Ismat Shah, S., Hasanain, S. K. *Journal of Magnetism and Magnetic Materials*, **280**(2–3), 412–418 (2004).
5. Kechrakos, D., and Trohidou, K. N. *Applied Surface Science.* **226**, 261–264(2004).
6. Tackett, R., Sudakar, C., Naik, R., Lawes, G., Rablau, C., and Vaishnava, P. P. *Journal of Magnetism and Magnetic Materials*, **320**(21), 2755–2759 (2008).
7. Koichiro, Hayashi, Wataru, Sakamoto, and Toshinobu, Yogo. *Journal of Magnetism and Magnetic Materials*, **321**(5), 450–457 (March, 2009).
8. Gufei, Zhang, Potzger, K., Shengqiang, Zhou, Mücklich, A., and Yicong, Ma. *J. Nuclear Instruments and Methods in Physics Research Section B: Beam Interactions with Materials and Atoms*, **267**(8–9), 1596–1599 (2009).
9. Daiji, Hasegawa, Haitao, Yang, Tomoyuki, Ogawa, Migaku, and Takahashi. *Journal of Magnetism and Magnetic Materials*, **321**(7), 746–749 (2009).
10. Denkbas, E. B., Kilicay, E., Birlikseven, C., and Ozturk, E. *Reactive and Functional Polymers*, **50**, 225–232 (2002).
11. Hyeon, T. *Chem. Commun.*, p. 927–934 (2003).
12. Jun, Y., Choi, J., and Cheon. J. *Chem. Commun.*, p. 1203–1214 (2007).
13. Kawamura, M., and Sato, K. *Chem. Commun.*, p. 3404–3405 (2007).
14. Muller, J. L., Klankermayer, J., and Leitner, W., *Chem. Commun.*, p. 1939–1941 (2007).
15. Dvorikova, R. A., Antipov, B. G., Klemenkova, Z. S., Shanditsev, V. A., Prokof'ev, A. I., Petrovskii, P. V., Rusanov, A. L., and Korshak, Yu. V. *Polymer Science Ser, A.* **47**(11), 1135–1140 (2005).
16. Dvorikova, R. A., Nikitin, L. N., Korshak, Yu. V., Shanditsev, V. A., Rusanov, A. L., Abramchuk, S. S., and Khokhlov, A. *R.Doklady Akademii Nauk*, **422**(3), 334–338 (2013).
17. Cooper, A. I. *J. Mater. Chem*, **10**, 207–234 (2000).
18. Nalawade, S. P., Picchioni, F., and Janssen, L. P. B. M. *Prog. Polym. Sci*, **31**, 19–43 (2006).
19. Reverchon, E., and Adami, R. *J. of Supercritical Fluids*, **37**, 1–22 (2006).
20. Zhang, Y., and Erkey, C. *J. of Supercritical Fluids*, **38**, 252–267 (2006).
21. Erkey, C. *J. of Supercritical Fluids*, **47**, 517–522 (2009).
22. Yang, J., Hasell, T., Wang, W., Howdle, S. M. *European Polymer Journal*, **44**, 1331–1336 (2008).
23. Said-Galiyev Ernest, E., Vygodskii Yakov, S., Nikitin Lev, N., Vinokur Rostislav, A., Gallyamov Marat, O., Pototskaya Inna, V., Kireev Vyacheslav, V., Khokhlov. Alexei, R., and Schaumburg, Kjeld. *J. of Supercritical Fluids*, **27**, 121–130 (2003).

24. Nikitin Lev, N., Gallyamov Marat, O., Vinokur Rostislav, A., Nikolaev Alexander, Yu., Said-Galiyev Ernest ,E., Khokhlov Alexei, R., Jespersen Henrik, T., and Schaumburg, Kjeld. *J. of Supercritical Fluids*, **265**, 263–273 (2003).

25. Yuvaraj, H., Woo, M. H., Park, E. J., Jeong, Y. T., and Lim, K. T. *European Polymer Journal*, **44**, 637–644 (2008).

26. Chen, A. Z., Kang, Y. Q., Pu, X. M., Yin, G. F., Li, Y., and Hu, J. Y. Development of Fe_3O_4-poly(l-lactide) magnetic microparticles in supercritical CO_2. *Journal of Colloid and Interface Science*, **330**, 317–322 (2009).

27. Tsang, S. C., Yu, C. H., Gao, X., and Tam, K. Y. Preparation of nanomagnetic absorbent for partition coefficient measurement. *International Journal of Pharmaceutics.*, **327**, 139–144 (2006).

28. Said-Galiyev, E., Nikitin, L., Vinokur, R., Gallyamov, M., Kurykin, M., Petrova, O., Lokshin, B., Volkov, I., Khokhlov, A., and Schaumburg, K. New chelate complexes of copper and iron: synthesis and impregnation into a polymer matrix from solution in supercritical carbon dioxide. *Industrial and Engineering Chem. Research*, **39**, 4891–4896 (2000).

29. Blackburn, J. M., Long, D. L., Cabanas, A., Watkins, J. J. Deposition of Conformal Copper and Nickel Films from Supercritical Carbon Dioxide. *Science*, **294**, 141–145 (2001).

30. Vasilkov, A., Naumkin, A., Nikitin, L., Volkov, I., Podshibikhin, V., and Lisichkin, G. Ultrahigh molecular weight polyethylene modified with silver nanoparticles prepared by metal-vapour synthesis. *AIP Conference Proceedings*, **1042**, 255–257 (2008).

31. Nikitin, L., Vasilkov, A., Vopilov, Yu., Buzin, M., Abramchuk, S., Bouznik, V., and Khokhlov, A. Making of metal-polymeric composites. *AIP Conference Proceedings*, **1042**, 249–251 (2008).

32. Nikitin, L. N., Vasilkov, A. Yu., Naumkin, A. V., Khokhlov, A. R., and Bouznik, V. M. Metal-polymeric composites prepared by supercritical carbon dioxide treatment and metal-vapor synthesis in: Success in Chemistry and Biochemistry: Mind's Flight in Time and Space, Volume 4, G. E. Zaikov (Ed), Nova Science Publishers, Inc. New York, pp. 579–590 (2009).

33. Dvorikova, R. A., Nikitin, L. N., Korshak, Yu. V., Shanditsev, V. A., Rusanov, A. L., Abramchuk, S. S., and Khokhlov, A. R. *New magnetic nanomaterials of hyperbranched ferrocenecontaining polyphenylenes prepared in liquid and supercritical carbon dioxide in: Quantitative Foundation of Chemical Reactions*. G. E. Zaikov (N.M. Emanuel Institute of Biochemical Physics, Russian Academy of Sciences, Moscow, Russia), Tanislaw, Grzegosz (Kaminski Institute of Natural Fibres, Poland), and Lev N. Nikitin (A.N.Nesmeyanov Institute of Organoelement Compounds, Russia). Nova Science Publishers, Inc. New York., pp. 93–100 (2009).

34. Sasaki, Yu., and Pittman, Ch. U. Acid-Catalyzed Reaction of Acetylferrocene with Trietyl Orthoformate. *J.Org.Chem.*, **38**(21), 3723–3726 (1973).

35. Seyoum, H. M., Bennet, L. H., and Della, Torre E. Temporal and temperature variations of dc magnetic aftereffect measurements of Fe3O4 powders. *J Appl Phys.*, **5**, 2820–2822 (2003).

CHAPTER 8

NEW ISSUES ON APPLICATION OF METAL-ORGANIC FRAMEWORKS (MOFS) ON TEXTILES

M. HASANZADEH and B. HADAVI MOGHADAM

CONTENTS

8.1   INTRODUCTION

Recently the application of nanostructured materials has garnered attention, due to their interesting chemical and physical properties. Application of nanostructured materials on the solid substrate such as fibers brings new properties to the final textile product [1]. Metal-organic frameworks (MOFs) are one of the most recognized nanoporous materials, which can be widely used for modification of fibers. These relatively crystalline materials consist of metal ions or clusters (named secondary building units, SBUs) interconnected by organic molecules called linkers, which can possess one, two or three dimensional structures [2-10]. They have received a great deal of attention, and the increase in the number of publications related to MOFs in the past decade is remarkable (Figure 1).

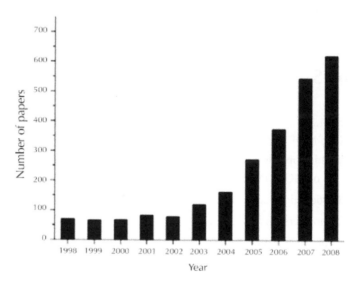

FIGURE 1   Number of publications on MOFs over the past decade, showing the increasing research interest in this topic

These materials possess a wide array of potential applications in many scientific and industrial fields, including gas storage [11,12], molecular separation [13], catalysis [14], drug delivery [15], sensing [16], and others. This is due to the unique combination of high porosity, very large surface areas, accessible pore volume, wide range of pore sizes and topologies, chemical stability, and infinite number of possible structures [17,18].

Although other well-known solid materials such as zeolites and active carbon also show large surface area and nanoporosity, MOFs have some new and distinct advantages. The most basic difference of MOFs and their inorganic counterparts (e.g. zeolites) is the chemical composition and absence of an inaccessible volume (called dead volume) in MOFs [10]. This feature offers the highest value of surface area and porosities in MOFs materials [19]. Another difference between MOFs and other well-known nanoporous materials such as zeolites and carbon nanotubes have the ability to tune the structure and functionality of MOFs directly during synthesis [17].

The first report of MOFs dates back to 1990, when Robson introduced a design concept to the construction of 3D MOFs using appropriate molecular building blocks and metal ions. Following the seminal work, several experiments were developed in this field such as work from Yaghi and O'Keeffe [20].

In this review, synthesis and structural properties of MOFs are summarized and some of the key advances that have been made in the application of these nanoporous materials in textile fibers are highlighted.

8.2 SYNTHESIS OF MOFS

The MOFs are typically synthesized under mild temperature (up to 200°C) by combination of organic linkers and metal ions (Figure 2) in solvothermal reaction [2,21].

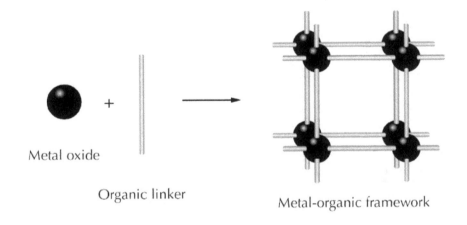

Metal oxide

Organic linker

Metal-organic framework

FIGURE 2   Formation of metal organic frameworks.

Recent studies have shown that the character of the MOF depends on many parameters including characteristics of the ligand (bond angles, ligand length, bulkiness, chirality, etc.), solubility of the reactants in the solvent, concentration of organic link and metal salt, solvent polarity, the pH of solution, ionic strength of the medium, temperature and pressure [2,21].

In addition to this synthesis method, several different methodologies are described in the literature such as ball-milling technique, microwave irradiation, and ultrasonic approach [22].

Post-synthetic modification (PSM) of MOFs opens up further chemical reactions to decorate the frameworks with molecules or functional groups that might not be achieved by conventional synthesis. In situations that presence of a certain functional group on a ligand prevents the formation of the targeted MOF, it is necessary to first form a MOF with the desired topology, and then add the functional group to the framework [2].

8.3   STRUCTURE AND PROPERTIES OF MOFS

When considering the structure of MOFs, it is useful to recognize the secondary building units (SBUs), for understanding and predicting topologies of structures [3]. Figure 3 shows the examples of some SBUs that are commonly occurring in metal carboxylate MOFs. Figure 3(a-c) illustrates inorganic SBUs include the square paddlewheel, the octahedral basic zinc acetate cluster, and the trigonal prismatic oxo-centered trimer, respectively. These SBUs are usually reticulated into MOFs by linking the carboxylate carbons with organic units [3]. Examples of organic SBUs are also shown in Figure 3(d–f).

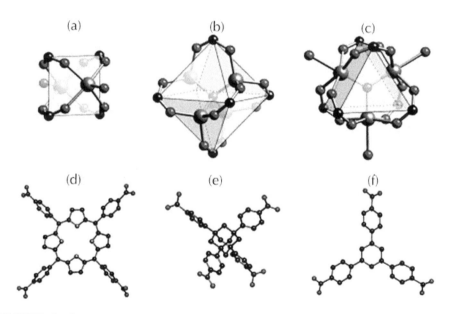

FIGURE 3   Structural representations of some SBUs, including (a-c) inorganic, and (b-f) organic SBUs. (Metals are shown as blue spheres, carbon as black spheres, oxygen as red spheres, nitrogen as green spheres)

It should be noted that the geometry of the SBU is dependent on not only the structure of the ligand and type of metal utilized, but also the metal to ligand ratio, the solvent, and the source of anions to balance the charge of the metal ion [2].

A large number of MOFs have been synthesized and reported by researchers to date. Isoreticular metal-organic frameworks (IRMOFs) denoted as IRMOF-n (n = 1 through 7, 8, 10, 12, 14, and 16) are one of the most widely studied MOFs in the literature. These compounds possess cubic framework structures in which each member shares the same cubic topology [3,21]. Figure 4 shows the structure of IRMOF-1(MOF-5) as simplest member of IRMOF series.

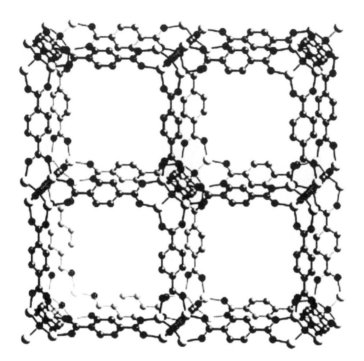

FIGURE 4 Structural representation of IRMOF-1. (Yellow, gray, and red spheres represent Zn, C, and O atoms, respectively).

8.4 APPLICATION OF MOFS IN TEXTILES

8.4.1 INTRODUCTION

There are many methods of surface modification, among which nanostructure based modifications have created a new approach for many applications in recent years. Although MOFs are one of the most promising nanostructured materials for modification of textile fibers, only a few examples have been reported to data. In this section, the first part focuses on application of MOFs in nanofibers and the second part is concerned with modifications of ordinary textile fiber with these nanoporous materials.

8.4.2 NANOFIBERS

Nanofibrous materials can be made by using the electrospinning process. Electrospinning process involves three main components including syringe filled with a polymer solution, a high voltage supplier to provide the required electric force for stretching the liquid jet, and a grounded collection plate to hold the nanofiber mat. The charged polymer solution forms a liquid jet that is drawn towards a grounded collection plate. During the jet movement to the collector, the solvent evaporates and dry fibers deposited as randomly oriented structure on the surface of a collector [23-28]. The schematic illustration of conventional electrospinning setup is shown in Figure 5.

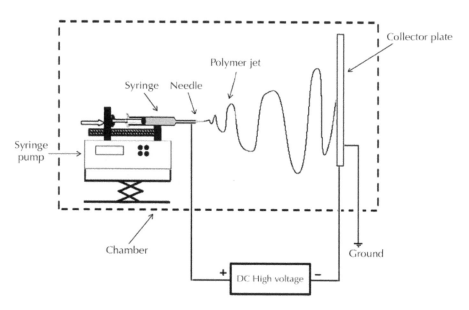

FIGURE 5   Schematic illustration of electrospinning set u.

At the present time, synthesis and fabrication of functional nanofibers represent one of the most interesting fields of nanoresearch. Combining the advanced structural features of metal-organic frameworks with the fabrication technique may generate new functionalized nanofibers for more multiple purposes.

While there has been great interest in the preparation of nanofibers, the studies on metal-organic polymers are rare. In the most recent investigation in this field, the growth of MOF (MIL-47) on electrospun polyacrylonitrile (PAN) mat was studied using in situ microwave irradiation [18]. MIL-47 consists of vanadium cations associated to six oxygen atoms, forming chains connected by terephthalate linkers (Figure 6).

It should be mentioned that the conversion of nitrile to carboxylic acid groups is necessary for the MOF growth on the PAN nanofibers surface.

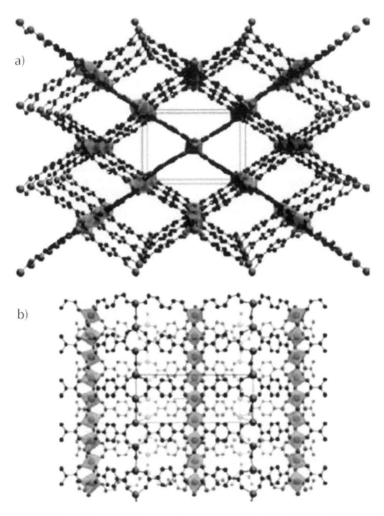

FIGURE 6   MIL-47 metal-organic framework structure: view along the b axis (a) and along the c axis (b

The crystal morphology of MIL-47 grown on the electrospun fibers illustrated that after only 5 s, the polymer surface was partially covered with small agglomerates of MOF particles. With increasing irradiation time, the agglomerates grew as elongated anisotropic structures (Figure 7) [18].

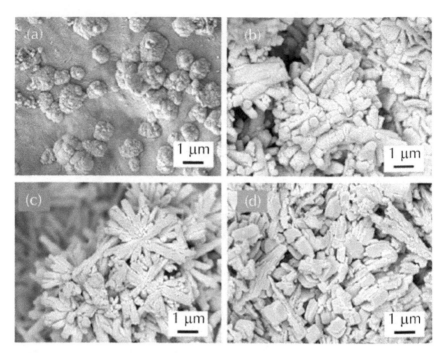

FIGURE 7   SEM micrograph of MIL-47 coated PAN substrate prepared from electrospun nanofibers as a function of irradiation time: (a) 5 s, (b) 30 s, (c) 3 min, and (d) 6 min

It is known that the synthesis of desirable metal-organic polymers is one of the most important factors for the success of the fabrication of metal-organic nanofibers [29]. Among several novel microporous metal-organic polymers, only a few of them have been fabricated into metal-organic fibers.

For example, new acentric metal-organic framework was synthesized and fabricated into nanofibers using electrospinning process [29]. The two dimensional network structure of synthesized MOF is shown in Figure 8. For this purpose, MOF was dissolved in water or DMF and saturated MOF solution was used for electrospinning. They studied the diameter and morphology of the nanofibers using an optical microscope and a scanning electron microscope (Figure 9). This fiber display diameters range from 60 nm to 4 μm.

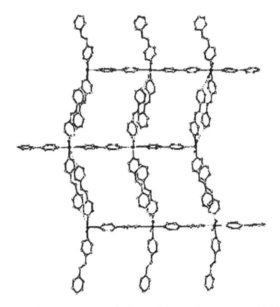

FIGURE 8   Representation of polymer chains and network structure of MO.

FIGURE 9   SEM micrograph of electrospun nanofibe.

In 2011, Kaskel et al. [30], reported the use of electrospinning process for the immobilization of MOF particles in fibers. They used HKUST-1 and MIL-100(Fe) as MOF particles, which are stable during the electrospinning process from a suspension. Electrospun polymer fibers containing up to 80 wt% MOF particles were achieved and exhibit a total accessible inner surface area. It was found that HKUST-1/PAN gives a spider web-like network of the fibers with MOF particles like trapped flies in it, while

HKUST-1/PS results in a pearl necklace-like alignment of the crystallites on the fibers with relatively low loadings.

8.4.3 ORDINARY TEXTILE FIBERS

Some examples of modification of fibers with metal-organic frameworks have verified successful. For instance, in the study on the growth of $Cu_3(BTC)_2$ (also known as HKUST-1, BTC=1,3,5-benzenetricarboxylate) MOF nanostructure on silk fiber under ultrasound irradiation, it was demonstrated that the silk fibers containing $Cu_3(BTC)_2$ MOF exhibited high antibacterial activity against the gram-negative bacterial strain *E. coli* and the gram-positive strain *S. aureus* [1]. The structure and SEM micrograph of $Cu_3(BTC)_2$ MOF is shown in Figure 10.

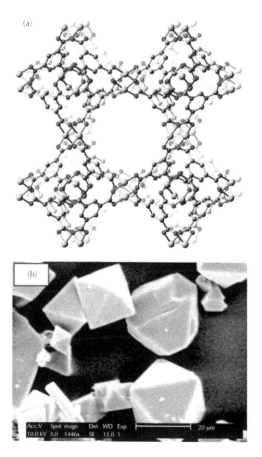

FIGURE 10   (a) The unit cell structure and (b) SEM micrograph of the $Cu_3(BTC)_2$ metal-organic framework. (Green, gray, and red spheres represent Cu, C, and O atoms, respectively)

$Cu_3(BTC_2$ MOF has a large pore volume, between 62% and 72% of the total volume, and a cubic structure consists of three mutually perpendicular channels [32].

The formation mechanism of $Cu_3(BTC)_2$ nanoparticles upon silk fiber is illustrated in Figure 11. It is found that formation of $Cu_3(BTC)_2$ MOF on silk fiber surface was increased in presence of ultrasound irradiation. In addition, increasing the concentration cause an increase in antimicrobial activity [1]. Figure 12 shows the SEM micrograph of $Cu_3(BTC)_2$ MOF on silk surface.

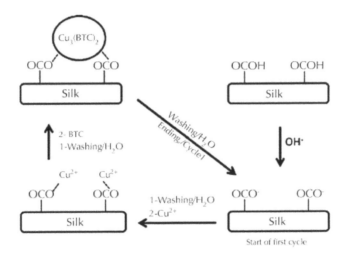

FIGURE 11   Schematic representation of the formation mechanism of $Cu_3(BTC)_2$ nanoparticles upon silk fibe.

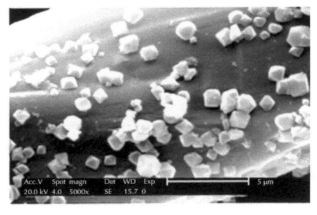

FIGURE 12   SEM micrograph of $Cu_3(BTC)_2$ crystals on silk fiber.

The FT-IR spectra of the pure silk yarn and silk yarn containing MOF (CuBTC-Silk) are shown in Figure 13. Owing to the reduction of the C=O bond, which is caused by the coordination of oxygen to the Cu^{2+} metal center (Figure 11), the stretching frequency of the C=O bond was shifted to lower wavenumbers (1654 cm⁻¹) in comparison with the free silk (1664 cm⁻¹) after chelation [1].

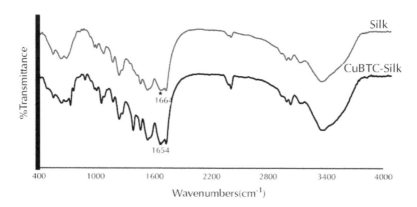

FIGURE 13   FT-IR spectra of the pure silk yarn and silk yarn containing $Cu_3(BTC)_2$

In another study, $Cu_3(BTC)_2$ was synthesized in the presence of pulp fibers of different qualities [33]. The following pulp samples were used: a bleached and an unbleached kraft pulp, and chemithermomechanical pulp (CTMP).

All three samples differed in their residual lignin content. Indeed, owing to the different chemical composition of samples, different results regarding the degree of coverage were expected. The content of $Cu_3(BTC)_2$ in pulp samples, k-number, and single point BET surface area are shown in Table 1. k-number of pulp samples, which is indicates the lignin content indirectly, was determined by consumption of a sulfuric permanganate solution of the selected pulp sample [33].

TABLE 1   Some characteristics of the pulp sampls

Pulp sample	MOF content[a] (wt.%)	**k-number**[b]	**Surface area**[c] **(m² g⁻¹)**
CTMP	19.95	114.5	314
Unbleached kraft pulp	10.69	27.6	165
Bleached kraft pulp	0	0.3	10

[a] Determined by thermogravimetric analysis.
[b] Determined according to ISO 302.
[c] Single point BET surface area calculated at p/p0=0.3 bar.

It is found that CTMP fibers showed the highest lignin residue and largest BET surface area. As shown in the SEM micrograph (Figure 14), the crystals are regularly distributed on the fiber surface. The unbleached kraft pulp sample provides a slightly lower content of MOF crystals and BET surface area with 165 $m^2 g^{-1}$. Moreover, no crystals adhered to the bleached kraft pulp, which was almost free of any lignin.

FIGURE 14   SEM micrograph of $Cu_3(BTC)_2$ crystals on the CTMP fibers

8.5   CONCLUSION

New review on feasibility and application of several kinds of metal-organic frameworks on different substrate including nanofiber and ordinary fiber was investigated. Based on the researcher's results, the following conclusions can be drawn:

1. Metal-organic frameworks (MOF), as new class of nanoporous materials, can be used for modification of textile fibers.
2. These nanostructured materials have many exciting characteristics such as large pore sizes, high porosity, high surface areas, and wide range of pore sizes and topologies.
3. Although tremendous progress has been made in the potential applications of MOFs during past decade, only a few investigations have reported in textile engineering fields.
4. Morphological properties of the MOF/fiber composites were defined; the most advantageous, particle size distribution was shown.
5. It is concluded that the MOFs/fiber composite would be good candidates for many technological applications, such as gas separation, hydrogen storage, sensor, and others.

KEYWORDS

- **Electrospun Polymer Fibers**
- **Nanoporous Materials**
- **Solvothermal Reaction**
- **Electrospinning Process**

REFERENCES

1. Abbasi, A. R., Akhbari, K., and Morsali, A. Dense coating of surface mounted CuBTC metal-organic framework nanostructures on silk fibers, prepared by layer-by-layer method under ultrasound irradiation with antibacterial activity. *Ultrasonics Sonochemistry*, **19**, 846-852 (2012).
2. Kuppler, R. J., Timmons, D. J., Fang, Q. R., Li, J. R., Makal, T. A.,Young, M. D., Yuan, D., Zhao, D., Zhuang, W., and Zhou, H. C. Potential applications of metal-organic frameworks. *Coordination Chemistry Reviews*, **253**, 3042-3066 (2009).
3. Rowsell, J. L. C., and Yaghi, O. M. Metal-organic frameworks A new class of porous materials. *Microporous and Mesoporous Materials*, **73**, 3-14 (2004).
4. An, J., Farha, O. K., Hupp, J. T., Pohl, E., Yeh, J. I., and Rosi, N. L. Metal-adeninate vertices for the construction of an exceptionally porous metal-organic framework. *Nature communications*, DOI: 10.1038/ncomms 1618, (2012).
5. Morris, W., Taylor, R. E., Dybowski, C., Yaghi, O. M., and Garcia-Garibay, M. A. Framework mobility in the metal-organic framework crystal IRMOF-3 Evidence for aromatic ring and amine rotation. *Journal of Molecular Structure*, **1004**,94-101 (2011).
6. Kepert, C. J. Metal-organic framework materials. in 'Porous Materials'. D. W. Bruce, D. O'Hare, and R. I. Walton (Eds). John Wiley & Sons, Chichester (2011).
7. Rowsell, J. L. C., and Yaghi, O. M. Effects of functionalization, catenation, and variation of the metal oxide and organic linking units on the low-pressure hydrogen adsorption properties of metal-organic frameworks. *Journal of the American Chemical Society*, **128**, 1304-1315 (2006).
8. Rowsell, J. L. C., and Yaghi, O. M. Strategies for hydrogen storage in metal-organic frameworks. *Angewandte Chemie International Edition*, **44**, 4670-4679 (2005).
9. Farha, O. K., Mulfort, K. L., Thorsness, A. M., and Hupp, J. T. Separating solids purification of metal-organic framework materials. *Journal of the American Chemical Society*, **130**, 8598-8599 (2008).
10. Khoshaman, A. H. Application of electrospun thin films for supra-molecule based gas sensing. M.Sc. thesis, Simon Fraser University (2011).
11. Murray, L. J., Dinca, M., and Long, J. R. Hydrogen storage in metal-organic frameworks. *Chemical Society Reviews*, **38**, 1294-1314 (2009).
12. Collins, D. J., and Zhou, H. C. Hydrogen storage in metal-organic frameworks. *Journal of Materials Chemistry*, **17**, 3154-3160 (2007).
13. Chen, B., Liang, C., Yang, J., Contreras, D. S., Clancy, Y. L., Lobkovsky, E. B., Yaghi, O. M., and Dai, S. A microporous metal-organic framework for gas-chromatographic separation of alkanes. *Angewandte Chemie International Edition*, **45**, 1390-1393 (2006).
14. Lee, J. Y., Farha, O. K., Roberts, J., Scheidt, K. A., Nguyen, S. T., and Hupp, J. T. Metal-organic framework materials as catalysts. *Chemical Society Reviews*, **38**, 1450-1459 (2009).

15. Huxford, R. C., Rocca, J. D., and Lin, W. Metal-organic frameworks as potential drug carriers. *Current Opinion in Chemical Biology*, **14**, 262-268 (2010).
16. Suh, M. P., Cheon, Y. E., and Lee, E. Y. Syntheses and functions of porous metallosupramolecular networks. *Coordination Chemistry Reviews*, **252**, 1007-1026 (2008).
17. Keskin, S., and Kızılel S. Biomedical applications of metal organic frameworks. *Industrial and Engineering Chemistry Research*. **50**, 1799-1812 (2011).
18. Centrone, A., Yang, Y., Speakman, S., Bromberg, L., Rutledge, G. C., and Hatton, T. A. Growth of metal-organic frameworks on polymer surfaces. *Journal of the American Chemical Society*, **132**, 15687-15691 (2010).
19. Wong-Foy, A. G., Matzger, A. J., and Yaghi, O. M. Exceptional H_2 saturation uptake in microporous metal-organic frameworks. *Journal of the American Chemical Society*, **128**, 3494-3495 (2006).
20. Farrusseng, D. Metal-organic frameworks Applications from Catalysis to Gas Storage. Wiley-VCH, Weinheim (2011).
21. Rosi, N. L., Eddaoudi, M., Kim, J., O'Keeffe, M., and Yaghi, O. M. Advances in the chemistry of metal-organic frameworks. *CrystEngComm*, 4, 401-404 (2002).
22. Zou, R., Abdel-Fattah, A. I., Xu, H., Zhao, Y., and Hickmott, D. D. Storage and separation applications of nanoporous metal-organic frameworks, *CrystEngComm*, **12**, 1337-1353 (2010).
23. Reneker, D. H., and Chun, I. Nanometer diameter fibers of polymer, produced by electrospinning, *Nanotechnology*, **7**, 216-223 (1996).
24. Shin, Y. M., Hohman, M. M., Brenner, M. P., and Rutledge, G. C. Experimental characterization of electrospinning, The electrically forced jet and instabilities. *Polymer,* **42**, 9955-9967 (2001).
25. Reneker, D. H., Yarin, A. L., Fong, H., and Koombhongse, S. Bending instability of electrically charged liquid jets of polymer solutions in electrospinning, *Journal of Applied Physics,* **87**, 4531-4547 (2000).
26. Zhang S., Shim W.S., Kim J.: Design of ultra-fine nonwovens via electrospinning of Nylon 6: Spinning parameters and filtration efficiency, *Materials and Design,* **30**, 3659-3666 (2009).
27. Yördem, O. S., Papila, M., and Menceloğlu, Y. Z. Effects of electrospinning parameters on polyacrylonitrile nanofiber diameter: An investigation by response surface methodology. *Materials and Design,* **29**, 34-44 (2008).
28. Chronakis, I. S. Novel nanocomposites and nanoceramics based on polymer nanofibers using electrospinning process-A review. *Journal of Materials Processing Technology*, **167**, 283-293 (2005).
29. Lu, J. Y., Runnels, K. A., and Norman, C. A new metal-organic polymer with large grid acentric structure created by unbalanced inclusion species and its electrospun nanofibers. *Inorganic Chemistry*, **40**, 4516-4517 (2001).
30. Rose, M., Böhringer, B., Jolly, M., Fischer, R., and Kaskel, S. MOF processing by electrospinning for functional textiles. *Advanced Engineering Materials*, **13**, 356-360 (2011).
31. Basu, S., Maes, M., Cano-Odena, A., Alaerts, L., De Vos, D. E., and Vankelecom, I. F. J. Solvent resistant nanofiltration (SRNF) membranes based on metal-organic frameworks. *Journal of Membrane Science*, **344**, 190-198 (2009).
32. Hopkins, J. B. Infrared spectroscopy of H_2 trapped in metal organic frameworks B.A. Thesis, Oberlin College Honors (2009).
33. [33] Küsgens, P., Siegle, S., and Kaskel, S. Crystal growth of the metal-organic framework $Cu_3(BTC)_2$ on the surface of pulp fibers. *Advanced Engineering Materials*, **11**, 93-95 (2009).

CHAPTER 9

A NEW APPROACH FOR OPTIMIZATION OF ELECTROSPUN PAN NANOFIBER DIAMETER AND CONTACT ANGLE

MAHDI HASANZADEH, BENTOLHODA HADAVI MOGHADAM, and MOHAMMAD HASANZADEH MOGHADAM ABATARI

CONTENTS

9.1 INTRODUCTION

Recently, it was demonstrated that electrospinning can produce superfine fiber ranging from micrometer to nanometer using an electric field force. In the electrospinning process, a strong electric field is applied between polymer solution contained in a syringe with a capillary tip and grounded collector. When the electric field overcomes the surface tension force, the charged polymer solution forms a liquid jet and travels towards collection plate. As the jet travels through the air, the solvent evaporates and dry fibers deposits on the surface of a collector [1-4].

The electrospun nanofibers have high specific surface area, high porosity, and small pore size. Therefore, they have been suggested as excellent candidate for many applications including filtration, multifunctional membranes, tissue engineering, protective clothing, reinforced composites, and hydrogen storage [5,6].

Studies have shown that the morphology and the properties of the electrospun nanofibers depend on many parameters including polymer solution properties (the concentration, liquid viscosity, surface tension, and dielectric properties of the polymer solution), processing parameters (applied voltage, volume flow rate, tip to collector distance, and the strength of the applied electric field), and ambient conditions (temperature, atmospheric pressure and humidity) [5-8].

Response surface methodology (RSM) is a combination of mathematical and statistical techniques used to evaluate the relationship between a set of controllable experimental factors and observed results. This optimization process is used in situations where several input variables influence some output variables of the system. The main goal of RSM is to optimize the response, which is influenced by several independent variables, with minimum number of experiments [9,10]. Therefore, the application of RSM in electrospinning process will be helpful in effort to find and optimize the electrospun nanofibers properties.

In this paper, a study has been conducted to investigate the relationship between four electrospinning parameters (solution concentration, applied voltage, tip to collector distance, and volume flow rate) and electrospun PAN nanofiber mat properties such as average fiber diameter (AFD) and contact angle (CA).

9.2 EXPERIMENTAL DETAILS

9.2.1 MATERIALS

Polyacrylonirile (PAN, M_w=100,000) was purchased from Polyacryle Co. (Iran) and N-N, dimethylformamide (DMF) was obtained from Merck Co. (Germany).

The polymer solutions with different concentration ranged from 10 wt.% to 14 wt.% were prepared by dissolving PAN powder in DMF and was stirred for 24 h at 50°C. These polymer solutions were used for electrospinning.

9.2.2 ELECTROSPINNING

A schematic of the electrospinning apparatus is shown in Figure 1. A polymer solution was loaded in a 5 mL syringe connected to a syringe pump. The tip of the syringe was connected to a high voltage power supply (capable to produce 0-40 kV). Under high voltage, a fluid jet was ejected from the tip of the needle and accelerated toward the grounded collector (aluminum foil). All electrospinnings were carried out at room temperature.

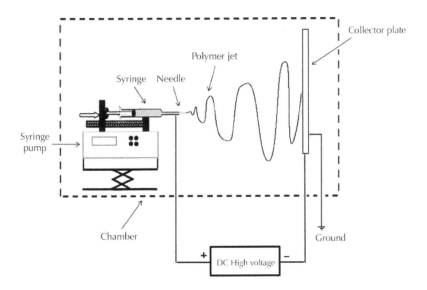

FIGURE 1   Schematic of electrospinning set up.

9.2.3   MEASUREMENT AND CHARACTERIZATION

The electrospun nanofibers were sputter-coated with gold and their morphology was examined with a scanning electron microscope (SEM, Philips XL-30). Average diameter of electrospun nanofibers was determined from selected SEM image by measuring at least 50 random fibers using Image J software.

The wettability of electrospun fiber mat was determined by water contact angle measurement. Contact angles were measured by specially arranged microscope equipped with camera and PCTV vision software as shown in Figure 2. The volume of the distilled water for each measurement was kept at 1 µl.

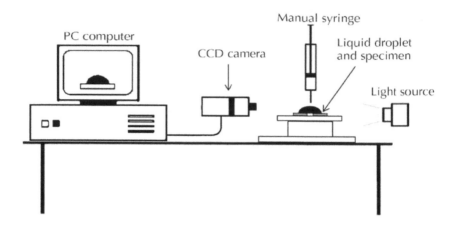

FIGURE 2 Schematic of contact angle measurement set up.

9.2.4 EXPERIMENTAL DESIGN BY RSM

In this study, the effect of four electrospinning parameters on two responses, comprising the AFD and the CA of electrospun fiber mat, was evaluated using central composite design (CCD). The experiment was performed for at least three levels of each factor to fit a quadratic model. Polymer solution concentration (X_1), applied voltage (X_2), tip to collector distance (X_3), and volume flow rate (X_4) were chosen as independent variables and the AFD and the CA of electrospun fiber mat as dependent variables (responses). The experimental parameters and their levels are given in Table 1.

TABLE 1 Design of experiment (factors and levels)

Factor	Variable	Unit	Factor level		
			-1	0	1
X_1	Solution concentration	(wt.%)	10	12	14
X_2	Applied voltage	(kV)	14	18	22
X_3	Tip to collector distance	(cm)	10	15	20
X_4	Volume flow rate	(ml/h)	2	2.5	3

A quadratic model, which also includes the linear model, is given below:

$$Y = \beta_0 + \sum_{i=1}^{4} \beta_i x_i + \sum_{i=1}^{4} \beta_{ii} x_i^2 + \sum_{i=1}^{3} \sum_{j=2}^{4} \beta_{ij} x_i x_j \tag{1}$$

where, Y is the predicted response, x_i and x_j are the independent variables, β_0 is a constant, β_i is the linear coefficients, β_{ii} is the squared coefficients and β_{ij} is the second-order interaction coefficients [9,10].

The statistical analysis of experimental data was performed using Design-Expert software (Version 8.0.3, Stat-Ease, Minneapolis, MN, 2010) including analysis of variance (ANOVA). A design of 30 experiments for independent variables and responses for AFD and CA are listed in Table 2.

TABLE 2   The actual design of experiments and responses for AFD and CA.

No.	Electrospinning parameters				Responses	
	X_1	X_2	X_3	X_4	AFD (nm)	CA (°)
	Concentration	Voltage	Distance	Flow rate		
1	10	14	10	2	206±33	44±6
2	10	22	10	2	187±50	54±7
3	10	14	20	2	162±25	61±6
4	10	22	20	2	164±51	65±4
5	10	14	10	3	225±41	38±5
6	10	22	10	3	196±53	49±4
7	10	14	20	3	181±43	51±5
8	10	22	20	3	170±50	56±5
9	10	18	15	2.5	188±49	48±3
10	12	14	15	2.5	210±31	30±3
11	12	22	15	2.5	184±47	35±5
12	12	18	10	2.5	214±38	22±3
13	12	18	20	2.5	205±31	30±4
14	12	18	15	2	195±47	33±4
15	12	18	15	3	221±23	25±3
16	12	18	15	2.5	199±50	26±4
17	12	18	15	2.5	205±31	29±3
18	12	18	15	2.5	225±38	28±5
19	12	18	15	2.5	221±23	25±4
20	12	18	15	2.5	215±35	24±3
21	12	18	15	2.5	218±30	21±3
22	14	14	10	2	255±38	31±4
23	14	22	10	2	213±37	35±5

TABLE 2   *(Continued)*

24	14	14	20	2	240±33	33±6
25	14	22	20	2	200±30	37±4
26	14	14	10	3	303±36	19±3
27	14	22	10	3	256±40	28±3
28	14	14	20	3	283±48	39±5
29	14	22	20	3	220±41	36±4
30	14	18	15	2.5	270±43	20±3

9.3   DISCUSSION AND RESULT

9.3.1   MORPHOLOGICAL ANALYSIS OF NANOFIBERS

The PAN solution in DMF were electrospun under different conditions, including various PAN solution concentrations, applied voltages, volume flow rates and tip to collector distances, to study the effect of electrospinning parameters on the morphology and properties of electrospun nanofibers.

Figure 3 shows the SEM images and fiber diameter distributions of electrospun fibers in different solution concentration as one of the most effective parameters to control the fiber morphology. As observed in Figure 3, the AFD increased with increasing solution concentration. It was suggested that the higher solution concentration would have more polymer chain entanglements and less chain mobility. This causes the hard jet extension and disruption during electrospinning process and producing thicker fibers.

(a)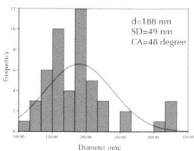

FIGURE 3   *(Continued)*

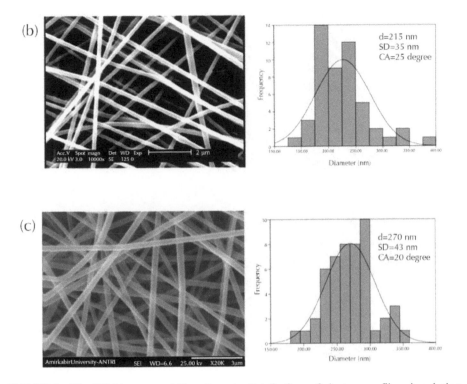

FIGURE 3   The SEM images and fiber diameter distributions of electrospun fibers in solution concentration of (a) 10 wt.%, (b) 12 wt.% and (c) 14 wt.%.

The SEM image and corresponding fiber diameter distribution of electrospun nanofiber in different applied voltage are shown in Figure 4. It is obvious that increasing the applied voltage cause an increase followed by a decrease in electrospun fiber diameter. As demonstrated by previous researchers [7,8], increasing the applied voltage may decrease, increase or may not change the fiber diameter. In one hand, increasing the applied voltage will increase the electric field strength and higher electrostatic repulsive force on the jet, favoring the thinner fiber formation. On the other hand, more surface charge will introduce on the jet and the solution will be removed more quickly from the tip of needle. As a result, the AFD will be increased [8,11].

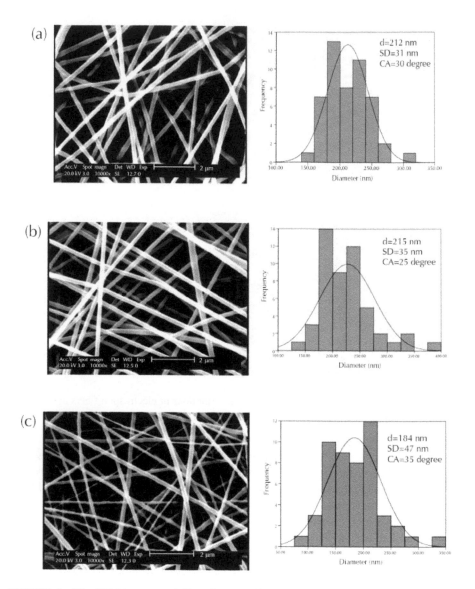

FIGURE 4 The SEM images and fiber diameter distributions of electrospun fibers in applied voltage of (a) 14 kV, (b) 18 kV and (c) 22 kV.

Figure 5 represents the SEM image and fiber diameter distribution of electrospun nanofiber in different spinning distance. It can be seen that the AFD decreased with increasing tip to collector distance. Because of the longer spinning distance could give more time for the solvent to evaporate, increasing the spinning distance will decrease fiber diameter [3,8].

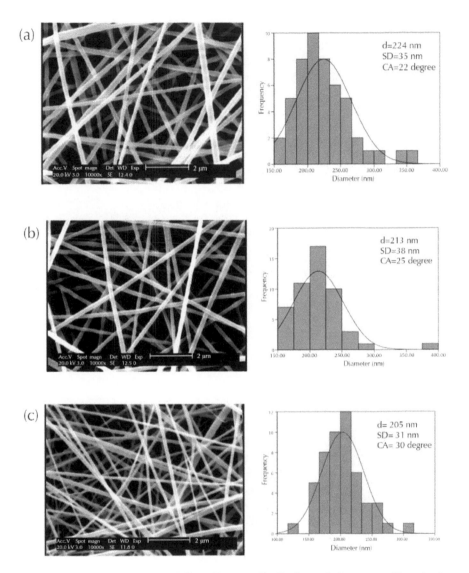

FIGURE 5   The SEM images and fiber diameter distributions of electrospun fibers in tip to collector distance of (a) 10 cm, (b) 15 cm and (c) 20 cm.

The SEM image and fiber diameter distribution of electrospun nanofiber in different volume flow rate are illustrated in Figure 6. It is clear that increasing the volume flow rate cause an increase in average fiber diameter. Ideally, the volume flow rate must be compatible with the amount of solution removed from the tip of the needle. At low volume flow rates, solvent would have sufficient time to evaporate and thinner fibers were produced, but at high volume flow rate, excess amount of solution fed to the tip of needle and thicker fibers result [3,12].

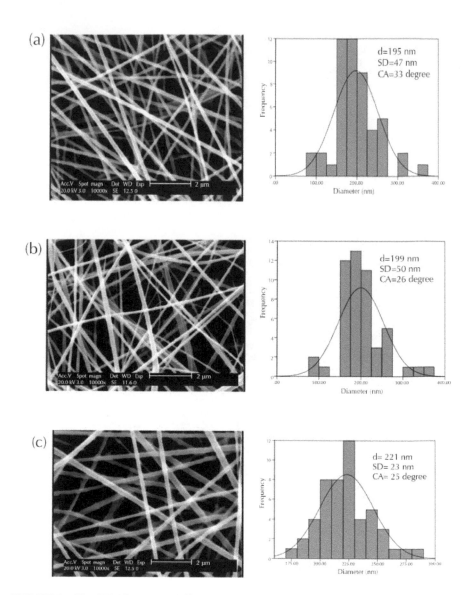

FIGURE 6   The SEM images and fiber diameter distributions of electrospun fibers in volume flow rate of (a) 2 ml/h, (b) 2.5 ml/h and (c) 3 ml/h.

9.3.2   THE ANALYSIS OF VARIANCE (ANOVA)

The analysis of variance for AFD and CA of electrospun fibers has been summarized in Table 3 and Table 4 respectively, which indicated that the predictability of the models is at 95% confidence interval. Using 5% significance level, the factor is considered significant if the p-value is less than 0.05.

From the p-values presented in Table 3 and Table 4, it is obvious that p-values of terms X_3^2, X_4^2, X_2X_3, X_1X_3, X_2X_4 and X_3X_4 in the model of AFD and X_3^2, X_4^2, X_2X_3, X_2X_4 and X_3X_4 in the model of CA were not significant (p>0.05).

The approximating function for AFD and CA of electrospun fiber obtained from Equation 2 and 3 respectively.

$$AFD = 211.89 + 31.17X_1 - 15.28X_2 - 12.78X_3 + 12.94X_4$$
$$- 8.44X_1X_2 + 6.31X_1X_4$$
$$+ 18.15X_1^2 - 13.85X_2^2$$

(2)

$$CA = 26.07 - 9.89X_1 + 2.17X_2 + 4.33X_3 - 2.33X_4$$
$$- 1.63X_1X_2 - 1.63X_1X_3 + 1.63X_1X_4$$
$$+ 9.08X_1^2 + 7.58X_2^2$$

(3)

TABLE 3   Analysis of variance for average fiber diameter (AFD).

Source	SS	DF	MS	F-value	Probe > F	Remarks
Model	31004.72	14	2214.62	28.67	<0.0001	Significant
X_1-Concentration	17484.50	1	17484.50	226.34	<0.0001	Significant
X_2-Voltage	4201.39	1	4201.39	54.39	<0.0001	Significant
X_3-Distance	2938.89	1	2938.89	38.04	<0.0001	Significant
X_4-Flow rate	3016.06	1	3016.06	39.04	<0.0001	Significant
X_1X_2	1139.06	1	1139.06	14.75	0.0016	Significant
X_1X_3	175.56	1	175.56	2.27	0.1524	
X_1X_4	637.56	1	637.56	8.25	0.0116	Significant
X_2X_3	39.06	1	39.06	0.51	0.4879	
X_2X_4	162.56	1	162.56	2.10	0.1675	
X_3X_4	60.06	1	60.06	0.78	0.3918	
X_1^2	945.71	1	945.71	12.24	0.0032	Significant
X_2^2	430.80	1	430.80	5.58	0.0322	Significant
X_3^2	0.40	1	0.40	0.0052	0.9433	

TABLE 3   *(Continued)*

X_4^2	9.30	1	9.30	0.12	0.7334
Residual	1158.75	15	77.25		
Lack of Fit	711.41	10	71.14	0.80	0.6468

TABLE 4   Analysis of variance for contact angle (CA) of electrospun fiber mat.

Source	SS	DF	MS	F-value	Probe > F	Remarks
Model	4175.07	14	298.22	32.70	<0.0001	Significant
X_1-Concentration	1760.22	1	1760.22	193.01	<0.0001	Significant
X_2-Voltage	84.50	1	84.50	9.27	0.0082	Significant
X_3-Distance	338.00	1	338.00	37.06	<0.0001	Significant
X_4-Flow rate	98.00	1	98.00	10.75	0.0051	Significant
X_1X_2	42.25	1	42.25	4.63	0.0481	Significant
X_1X_3	42.25	1	42.25	4.63	0.0481	Significant
X_1X_4	42.25	1	42.25	4.63	0.0481	Significant
X_2X_3	12.25	1	12.25	1.34	0.2646	
X_2X_4	6.25	1	6.25	0.69	0.4207	
X_3X_4	2.25	1	2.25	0.25	0.6266	
X_1^2	161.84	1	161.84	17.75	0.0008	Significant
X_2^2	106.24	1	106.24	11.65	0.0039	Significant
X_3^2	0.024	1	0.024	0.0026	0.9597	
X_4^2	21.84	1	21.84	2.40	0.1426	
Residual	136.80	15	9.12			
Lack of Fit	95.30	10	9.53	1.15	0.4668	

Analysis of variance for AFD and CA showed that the models were significant (p<0.0001), which indicated that the both models have a good agreement with experimental data. The value of determination coefficient (R^2) for AFD and CA was evaluated as 0.9640 and 0.9683 respectively.

The predicted versus actual response plots of AFD and CA are shown in Figures 7 and 8 respectively. It can be observed that experimental values are in good agreement with the predicted values.

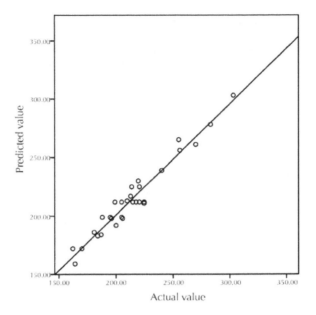

FIGURE 7   The predicted versus actual plot for AFD of electrospun fiber mat.

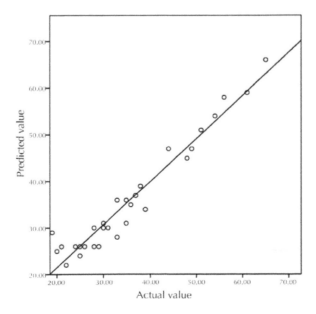

FIGURE 8   The predicted versus actual plot for CA of electrospun fiber mat.

9.3.3   EFFECTS OF SIGNIFICANT PARAMETERS ON AFD

The response surface and contour plots in Figure 9 (a) indicated that there was a considerable interaction between solution concentration and applied voltage at middle level of spinning distance (15 cm) and flow rate (2.5 ml/h). It can be seen an increase in AFD with increase in solution concentration at any given voltage that is in agreement with previous observations [11,12]. Generally, a minimum solution concentration is required to obtain uniform fibers from electrospinning. Below this concentration, polymer chain entanglements are insufficient and a mixture of beads and fibers is obtained. On the other hand, the higher solution concentration would have more polymer chain entanglements and less chain mobility. This causes the hard jet extension and disruption during electrospinning process and producing thicker fibers [7].

Figure 9 (b) shows the response surface and contour plots of interaction between solution concentration and flow rate at fixed voltage (18 kV) and spinning distance (15 cm). It can be seen that at fixed applied voltage and spinning distance, an increase in solution concentration and volume flow rate results in fiber with higher diameter. As mentioned in the literature, the volume flow rate must be compatible with the amount of solution removed from the tip of the needle. At low volume flow rates, solvent would have sufficient time to evaporate and thinner fibers were produced, but at high volume flow rate, excess amount of solution fed to the tip of needle and thicker fibers were resulted [3,8].

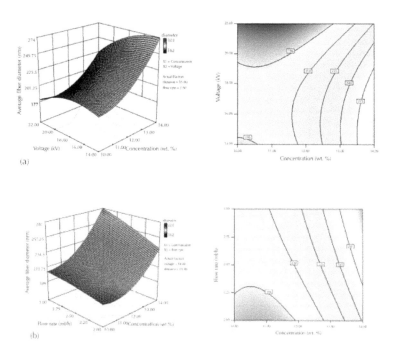

FIGURE 9   Response surface and contour plots of AFD showing the effect of: (a) solution concentration and applied voltage, (b) solution concentration and volume flow rate.

9.3.4 EFFECTS OF SIGNIFICANT PARAMETERS ON CA

The response surface and contour plots in Figure 10 (a) represented the CA of electrospun fiber mat at different solution concentration and applied voltage. It is obvious that at fixed spinning distance and volume flow rate, an increase in applied voltage and decrease in solution concentration result the higher CA. The tip to collector distance was found to be another important processing parameter as it influences the solvent evaporating rate and deposition time as well as electrostatic field strength. The impact of spinning distance on CA of electrospun fiber mat is illustrated in Figure 10 (b). Increasing the spinning distance causes the CA of electrospun fiber mat to increase. As demonstrated in Figure 10 (b), low solution concentration cause an increase in CA of electrospun fiber mat at large spinning distance. The response plots in Figure 10 (c) shows the interaction between solution concentration and volume flow rate at fixed applied voltage and spinning distance. It is obvious that at any given flow rate, CA of electrospun fiber mat will increase as solution concentration decreases.

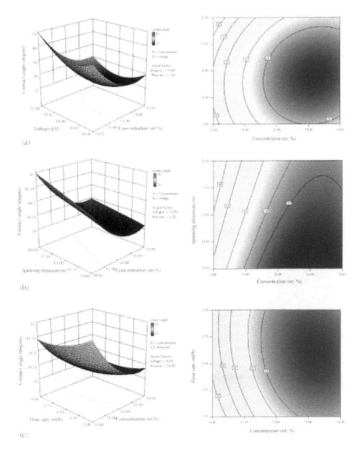

FIGURE 10   Response surface and contour plots of CA showing the effect of: (a) solution concentration and applied voltage, (b) solution concentration and spinning distance, (c) solution concentration and volume flow rate.

9.3.5   DETERMINATION OF OPTIMAL CONDITIONS

It is well known that the value of CA for hydrophilic surfaces is less than 90°. Fabrication of these surfaces has attracted considerable interest for both fundamental research and practical studies. So, the goal of the present study is to minimize the CA of electrospun nanofibers. The optimal conditions of the electrospinning parameters were established from the quadratic form of the RSM. Independent variables namely, solution concentration, applied voltage, spinning distance, and volume flow rate were set in range and dependent variable (CA) was fixed at minimum. The optimal conditions in the tested range for minimum CA of electrospun fiber mat are shown in Table 5.

This optimum condition was a predicted value, thus to confirm the predictive ability of the RSM model for response, a further electrospinning was carried out according to the optimized conditions and the agreement between predicted and measured responses was verified. The measured CA of electrospun nanofiber mat (21°) was very close to the predicted value estimated to 20°. Figure 11 shows the SEM image and AFD distribution of electrospun fiber mat prepared at optimized conditions.

TABLE 5   Optimum values of the process parameters for minimum CA of electrospun fiber mat.

Parameter	Optimum value
Solution concentration (wt.%)	13.2
Applied voltage (kV)	16.5
Spinning distance (cm)	10.6
Volume flow rate (ml/h)	2.5

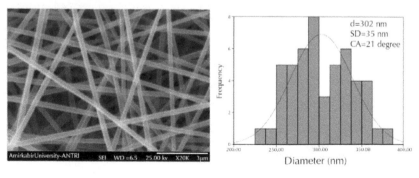

FIGURE 11   SEM image and fiber diameter distribution of electrospun fiber mat prepared at optimized conditions.

9.4 CONCLUSION

In this study, the effects of electrospinning parameters, comprising solution concentration (wt.%), applied voltage (kV), tip to collector distance (cm), and volume flow rate (ml/h) on average diameter and CA of electrospun PAN nanofibers were investigated by statistical approach. Response surface methodology (RSM) was successfully employed to model and optimize the electrospun nanofibers diameter and CA. The response surface and contour plots of the predicted AFD and CA indicated that the nanofiber diameter and its CA are very sensitive to solution concentration changes. It was concluded that the polymer solution concentration was the most significant factor impacting the AFD and CA of electrospun fiber mat. The R^2 value was 0.9640 and 0.9683 for AFD and CA respectively, which indicates a good fit of the models with experimental data. The optimum value of the solution concentration, applied voltage, spinning distance, and flow rate were found to be 13.2 wt%, 16.5 kV, 10.6 cm and 2.5 ml/h, respectively, for minimum CA of electrospun fiber mat.

KEYWORDS

- **Electrospinnings**
- **Electrospun Nanofibers**
- **Electrostatic Repulsive Force**

REFERENCES

1. Shams Nateri, A. and Hasanzadeh, M. J. Comput. Theor. Nanosci., 6, 1542 (2009).
2. Kilic, A., Oruc, F., and Demir, A. Text. Res. J., 78, 532 (2008).
3. Ramakrishna, S., Fujihara, K., Teo, W. E., Lim, T. C., and Ma, Z., Editors, An Introduction to Electrospinning and Nanofibers, National University of Singapore, World Scientific Publishing, Singapore (2005).
4. P. J. Brown and K. Stevens (Eds), Nanofibers and Nanotechnology in Textiles, Woodhead Publishing, Cambridge, UK (2007).
5. Shin, Y. M., Hohman, M. M., Brenner, M. P., and Rutledge, G. C. Polymer, 42, 9955 (2001).
6. Yördem, O. S., Papila, M., and Menceloğlu, Y. Z. Mater. Design, 29, 34 (2008).
7. Haghi, A. K. and Akbari, M. Phys. Status. Solidi. A., 204, 1830 (2007).
8. Ziabari, M., Mottaghitalab, V., and Haghi, A. K. in Nanofibers: Fabrication, Performance, and Applications, W. N. Chang (Ed.), Nova Science Publishers, USA, pp. 153–182 (2009),.
9. R. H. Myers, D. C. Montgomery, and C. M. Anderson-cook (Eds), Response surface methodology: process and product optimization using designed experiments, 3rd ed., John Wiley and Sons, USA (2009).
10. Gu, S. Y., Ren, J., and Vancso, G. J. Eur. Polym. J., 41, 2559 (2005).
11. Zhang, C., Yuan, X., Wu, L., Han, Y., and Sheng, J. Eur. Polym. J., 41, 423 (2005).
12. Zhang, S., Shim, W. S., and Kim, J. Mater. Design, 30, 3659 (2009).

CHAPTER 10

COMPARISON OF ANN WITH RSM IN PREDICTING CONTACT ANGLE OF ELECTROSPUN POLYACRYLONITRILE NANOFIBER MAT

B. HADAVI MOGHADAM and M. HASANZADEH

CONTENTS

10.1   INTRODUCTION

The wettability of solid surfaces is a very important property of surface chemistry, which is controlled by both the chemical composition and the geometrical microstructure of a rough surface [1-3]. When a liquid droplet contacts a rough surface, it will spread or remain as droplet with the formation of angle between the liquid and solid phases. Contact angle (CA) measurements are widely used to characterize the wettability of rough surface [3-5]. There are various methods to make a rough surface, such as electrospinning, electrochemical deposition, evaporation, chemical vapor deposition (CVD), plasma, and so on.

Electrospinning as a simple and effective method for preparation of nanofibrous materials have attracted increasing attention during the last two decade [6]. Electrospinning process, unlike the conventional fiber spinning systems (melt spinning, wet spinning, etc.), uses electric field force instead of mechanical force to draw and stretch a polymer jet [7]. This process involves three main components including syringe filled with a polymer solution, a high voltage supplier to provide the required electric force for stretching the liquid jet, and a grounded collection plate to hold the nanofiber mat. The charged polymer solution forms a liquid jet that is drawn towards a grounded collection plate. During the jet movement to the collector, the solvent evaporates and dry fibers deposited as randomly oriented structure on the surface of a collector [8-13]. The electrospun nanofiber mat possesses high specific surface area, high porosity, and small pore size. Therefore, they have been suggested as excellent candidate for many applications including filtration [14], multifunctional membranes [15], biomedical agents [16], tissue engineering scaffolds [17-18], wound dressings [19], full cell [20] and protective clothing [21].

The morphology and the CA of the electrospun nanofibers can be affected by many electrospinning parameters including solution properties (the concentration, liquid viscosity, surface tension, and dielectric properties of the polymer solution), processing conditions (applied voltage, volume flow rate, tip to collector distance, and the strength of the applied electric field), and ambient conditions (temperature, atmospheric pressure and humidity) [9-12].

In this work, the influence of four electrospinning parameters, comprising solution concentration, applied voltage, tip to collector distance, and volume flow rate, on the CA of the electrospun PAN nanofiber mat was carried out using response surface methodology (RSM) and artificial neural network (ANN). First, a central composite design (CCD) was used to evaluate main and combined effects of above parameters. Then, these independent parameters were fed as inputs to an ANN while the output of the network was the CA of electrospun fiber mat. Finally, the importance of each electrospinning parameters on the variation of CA of electrospun fiber mat was determined and comparison of predicted CA value using RSM and ANN are discussed.

10.2 EXPERIMENTAL

10.2.1 MATERIALS

PAN powder was purchased from Polyacryle Co. (Iran). The weight average molecular weight (M_w) of PAN was approximately 100,000 g/mol. *N-N*, dimethylformamide (DMF) was obtained from Merck Co. (Germany) and was used as a solvent. These chemicals were used as received.

10.2.2 ELECTROSPINNING

The PAN powder was dissolved in DMF and gently stirred for 24 hr at 50°C. Therefore, homogenous PAN/DMF solution was prepared in different concentration ranged from 10 wt.% to 14 wt.%. Electrospinning was set up in a horizontal configuration as shown in Figure 1. The electrospinning apparatus consisted of 5 ml plastic syringe connected to a syringe pump and a rectangular grounded collector (aluminum sheet). A high voltage power supply (capable to produce 0–40 kV) was used to apply a proper potential to the metal needle. It should be noted that all electrospinnings were carried out at room temperature.

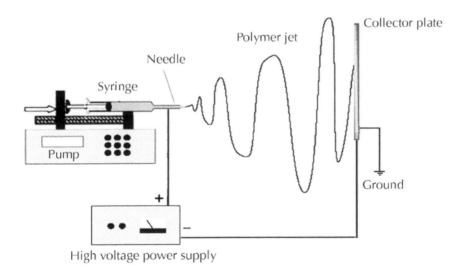

FIGURE 1   Schematic of electrospinning set up.

10.2.3 MEASUREMENT AND CHARACTERIZATION

The morphology of the gold-sputtered electrospun fibers were observed by scanning electron microscope (SEM, Philips XL-30). The average fiber diameter and distribution was determined from selected SEM image by measuring at least 50 random fibers. The wettability of electrospun fiber mat was determined by CA measurement. The CA measurements were carried out using specially arranged microscope equipped with camera and PCTV vision software as shown in Figure 2. The droplet used was distilled water and was 1 µl in volume. The CA experiments were carried out at room temperature and were repeated five times. All contact angles measured within 20 s of placement of the water droplet on the electrospun fiber mat. A typical SEM image of electrospun fiber mat, its corresponding diameter distribution and CA image are shown in Figure 3.

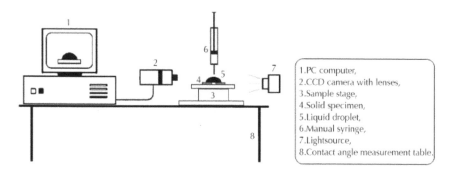

1.PC computer,
2.CCD camera with lenses,
3.Sample stage,
4.Solid specimen,
5.Liquid droplet,
6.Manual syringe,
7.Lightsource,
8.Contact angle measurement table.

FIGURE 2   Schematic of CA measurement set up.

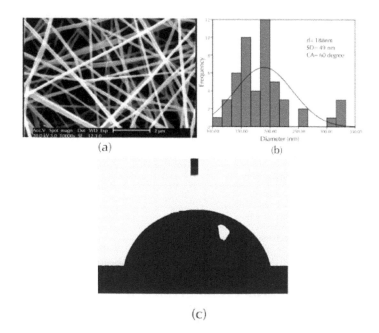

(a)

(b)

(c)

FIGURE 3   A typical (a) SEM image, (b) fiber diameter distribution, and (c) CA of electrospun fiber mat.

10.2.4 EXPERIMENTAL DESIGN

RESPONSE SURFACE METHODOLOGY

Response surface methodology (RSM) is a combination of mathematical and statistical techniques used to evaluate the relationship between a set of controllable experimental factors and observed results. This optimization process is used in situations where several input variables influence some output variables (responses) of the system [22-23].

In the present study, central composite design (CCD) was employed to establish relationships between four electrospinning parameters and the CA of electrospun fiber mat. The experiment was performed for at least three levels of each factor to fit a quadratic model. Based on preliminary experiments, polymer solution concentration (wt.%), applied voltage (kV), tip to collector distance (cm), and volume flow rate (ml/h) were determined as critical factors with significance effect on CA of electrospun fiber mat. These factors were four independent variables and chosen equally spaced, while the CA of electrospun fiber mat was dependent variable. The values of -1, 0, and 1 are coded variables corresponding to low, intermediate and high levels of each factor respectively. The experimental parameters and their levels for four independent variables are shown in Table 1. The regression analysis of the experimental data was carried out to obtain an empirical model between processing variables. The contour surface plots were obtained using Design-Expert® software.

TABLE 1   Design of experiment (factors and levels)

Factor	Variable	Unit	Factor level		
			-1	0	1
X_1	Solution concentration	(wt.%)	10	12	14
X_2	Applied voltage	(kV)	14	18	22
X_3	Tip to collector distance	(cm)	10	15	20
X_4	Volume flow rate	(ml/h)	2	2.5	3

The quadratic model, Equation (1) including the linear terms, was fitted to the data.

$$Y = \beta_0 + \sum_{i=1}^{4} \beta_i x_i + \sum_{i=1}^{4} \beta_{ii} x_i^2 + \sum_{i=1}^{3} \sum_{j=2}^{4} \beta_{ij} x_i x_j \qquad (1)$$

where, Y is the predicted response, x_i and x_j are the independent variables, β_0 is a constant, β_i is the linear coefficient, β_{ii} is the squared coefficient, and β_{ij} is the second-order interaction coefficients [22,23].

The quality of the fitted polynomial model was expressed by the determination coefficient (R^2) and its statistical significance was performed with the Fisher's statistical test for analysis of variance (ANOVA).

ARTIFICIAL NEURAL NETWORK

Artificial neural network (ANN) is an information processing technique, which is inspired by biological nervous system, composed of simple unit (neurons) operating in parallel (Figure 4). A typical ANN consists of three or more layers, comprising an input layer, one or more hidden layers and an output layer. Every neuron has connections with every neuron in both the previous and the following layer. The connections between neurons consist of weights and biases. The weights between the neurons play an important role during the training process. Each neuron in hidden layer and output layer has a transfer function to produce an estimate as target. The interconnection weights are adjusted, based on a comparison of the network output (predicted data) and the actual output (target), to minimize the error between the network output and the target [6,24-25].

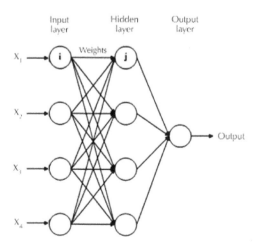

FIGURE 4   The topology of artificial neural network used in this study.

In this study, feed forward ANN with one hidden layer composed of four neurons was selected. The ANN was trained using back-propagation algorithm. The same experimental data used for each RSM designs were also used as the input variables of the ANN. There are four neurons in the input layer corresponding to four electrospinning parameters and one neuron in the output layer corresponding to CA of electrospun fiber mat. Figure 4 illustrates the topology of ANN used in this investigation.

10.3 DISCUSSION AND RESULTS

This section discusses in details the wettability behavior of electrospun fiber mat concluded from CA measurements. The results of the proposed RSM and ANN models are also presented followed by a comparison between those models.

10.3.1 THE ANALYSIS OF VARIANCE (ANOVA)

All 30 experimental runs of CCD were performed according to Table 2. A significance level of 5% was selected; that is, statistical conclusions may be assessed with 95% confidence. In this significance level, the factor has significant effect on CA if the p-value is less than 0.05. On the other hand, when p-value is greater than 0.05, it is concluded the factor has no significant effect on CA.

The results of analysis of variance (ANOVA) for the CA of electrospun fibers are shown in Table 3. Equation (2) is the calculated regression equation.

TABLE 2   The actual design of experiments and response

No.	X_1 Concentration	X_2 Voltage	X_3 Distance	X_4 Flow rate	CA (°)
	Electrospinning parameters				Response
1	10	14	10	2	446
2	10	22	10	2	54±7
3	10	14	20	2	61±6
4	10	22	20	2	65±4
5	10	14	10	3	38±5
6	10	22	10	3	49±4
7	10	14	20	3	51±5
8	10	22	20	3	56±5
9	10	18	15	2.5	48±3
10	12	14	15	2.5	30±3
11	12	22	15	2.5	35±5
12	12	18	10	2.5	22±3
13	12	18	20	2.5	30±4
14	12	18	15	2	33±4
15	12	18	15	3	25±3
16	12	18	15	2.5	26±4
17	12	18	15	2.5	29±3

TABLE 2 *(Continued)*

18	12	18	15	2.5	28±5
19	12	18	15	2.5	25±4
20	12	18	15	2.5	24±3
21	12	18	15	2.5	21±3
22	14	14	10	2	31±4
23	14	22	10	2	35±5
24	14	14	20	2	33±6
25	14	22	20	2	37±4
26	14	14	10	3	19±3
27	14	22	10	3	28±3
28	14	14	20	3	39±5
29	14	22	20	3	36±4
30	14	18	15	2.5	203

$$CA = 25.80 - 9.89X_1 + 2.17X_2 + 4.33X_3 - 2.33X_4$$
$$-1.63X_1X_2 - 1.63X_1X_3 + 1.63X_1X_4 - 0.88X_2X_3 - 0.63X_2X_4 + 0.37X_3X_4 \quad (2)$$
$$+7.90X_1^2 + 6.40X_2^2 - 0.096X_3^2 + 2.90X_4^2$$

TABLE 3 Analysis of variance for the CA of electrospun fiber mat

Source	SS	DF	MS	F-value	Probe > F	Remarks
Model	4175.07	14	298.22	32.70	<0.0001	Significant
X_1-Concentration	1760.22	1	1760.22	193.01	<0.0001	Significant
X_2-Voltage	84.50	1	84.50	9.27	0.0082	Significant
X_3-Distance	338.00	1	338.00	37.06	<0.0001	Significant
X_4-Flow rate	98.00	1	98.00	10.75	0.0051	Significant
X_1X_2	42.25	1	42.25	4.63	0.0481	Significant
X_1X_3	42.25	1	42.25	4.63	0.0481	Significant
X_1X_4	42.25	1	42.25	4.63	0.0481	Significant
X_2X_3	12.25	1	12.25	1.34	0.2646	
X_2X_4	6.25	1	6.25	0.69	0.4207	Significant
X_3X_4	2.25	1	2.25	0.25	0.6266	
X_1^2	161.84	1	161.84	17.75	0.0008	Significant
X_2^2	106.24	1	106.24	11.65	0.0039	Significant
X_3^2	0.024	1	0.024	0.0026	0.9597	
X_4^2	21.84	1	21.84	2.40	0.1426	
Residual	136.80	15	9.12			
Lack of Fit	95.30	10	9.53	1.15	0.4668	

From the p-values presented in Table 3, it can be concluded that the p-values of terms X_3^2, X_4^2, X_2X_3, X_2X_4 and X_3X_4 is greater than the significance level of 0.05, therefore they have no significant effect on the CA of electrospun fiber mat. Since the above terms had no significant effect on CA of electrospun fiber mat, these terms were removed. The fitted equations in coded unit are given in Equation (3).

$$
\begin{aligned}
CA = {} & 26.07 - 9.89X_1 + 2.17X_2 + 4.33X_3 - 2.33X_4 \\
& -1.63X_1X_2 - 1.63X_1X_3 + 1.63X_1X_4 \\
& +9.08X_1^2 + 7.58X_2^2
\end{aligned}
\tag{3}
$$

Now, all the p-values are less than the significance level of 0.05. Analysis of variance showed that the RSM model was significant ($p < 0.0001$), which indicated that the model has a good agreement with experimental data. The determination coefficient (R^2) obtained from regression equation was 0.958.

10.3.2 ARTIFICIAL NEURAL NETWORK

In this study, the best prediction, based on minimum error, was obtained by ANN with one hidden layer. The suitable number of neurons in the hidden layer was determined by changing the number of neurons. The good prediction and minimum error value were obtained with four neurons in the hidden layer. The weights and bias of ANN for CA of electrospun fiber mat are given in Table 4. The R^2 and mean absolute percentage error were 0.965 and 5.94% respectively, which indicates that the model was shows good fitting with experimental data.

TABLE 4   Weights and bias obtained in training ANN

		IW11	IW12	IW13	IW14
		1.0610	1.1064	21.4500	3.0700
		IW21	IW22	IW23	IW24
	Weights	-0.3346	2.0508	0.2210	-0.2224
		IW31	IW32	IW33	IW34
Hidden layer		-0.6369	-1.1086	-41.5559	0.0030
		IW41	IW42	IW43	IW44
		-0.5038	-0.0354	0.0521	0.9560
	Bias	b11	b21	b31	b41
		-2.5521	-2.0885	-0.0949	1.5478

TABLE 4　*(Continued)*

		LW11
		0.5658
		LW21
Output layer	Weights	0.2580
		LW31
		-0.2759
		LW41
		-0.6657
	Bias	b
		0.7104

10.3.3　EFFECTS OF SIGNIFICANT PARAMETERS ON RESPONSE

The morphology and structure of electrospun fiber mat, such as the nanoscale fibers and interfibrillar distance, increases the surface roughness as well as the fraction of contact area of droplet with the air trapped between fibers. It is proved that the CA decrease with increasing the fiber diameter [26], therefore the thinner fibers, due to their high surface roughness, have higher CA than the thicker fibers. Hence, we used this fact for comparing CA of electrospun fiber mat. The interaction contour plot for CA of electrospun PAN fiber mat are shown in Figure 5.

As mentioned in the literature, a minimum solution concentration is required to obtain uniform fibers from electrospinning. Below this concentration, polymer chain entanglements are insufficient and a mixture of beads and fibers is obtained. On the other hand, the higher solution concentration would have more polymer chain entanglements and less chain mobility. This causes the hard jet extension and disruption during electrospinning process and producing thicker fibers [27]. Figure 5 (a) show the effect of solution concentration and applied voltage at middle level of distance (15 cm) and flow rate (2.5 ml/h) on CA of electrospun fiber mat. It is obvious that at any given voltage, the CA of electrospun fiber mat decrease with increasing the solution concentration.

Figure 5 (b) shows the response contour plot of interaction between solution concentration and spinning distance at fixed voltage (18 kV) and flow rate (2.5 ml/h). Increasing the spinning distance causes the CA of electrospun fiber mat to increase. Because of the longer spinning distance could give more time for the solvent to evaporate, increasing the spinning distance will decrease the nanofiber diameter and increase the CA of electrospun fiber mat [28,29]. As demonstrated in Figure 5 (b), low solution concentration cause the increase in CA of electrospun fiber mat at large spinning distance.

The response contour plot in Figure 5 (c) represented the CA of electrospun fiber mat at different solution concentration and volume flow rate. Ideally, the volume flow rate must be compatible with the amount of solution removed from the tip of the needle. At low volume flow rates, solvent would have sufficient time to evaporate and thinner fibers were produced, but at high volume flow rate, excess amount of solution fed to the tip of needle and thicker fibers were resulted [28-30]. Therefore the CA of electrospun fiber mat will be decreased.

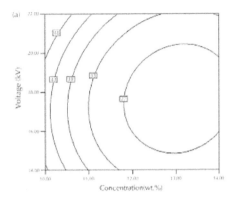

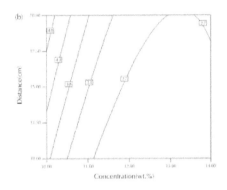

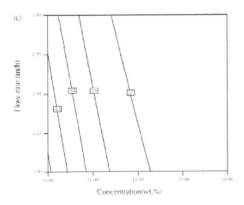

FIGURE 5   Contour plots for contact angle of electrospun fiber mat showing the effect of: (a) solution concentration and applied voltage, (b) solution concentration and spinning distance, (c) solution concentration and volume flow rate.

As shown by Equation (4), the relative importance (RI) of the various input variables on the output variable can be determined using ANN weight matrix [31].

$$RI_j = \frac{\sum_{m=1}^{N_h}\left(\left(\left|IW_{jm}\right|\Big/\sum_{k=1}^{N_i}\left|IW_{km}\right|\right)\times\left|LW_{mn}\right|\right)}{\sum_{k=1}^{N_i}\left\{\sum_{m=1}^{N_h}\left(\left(\left|IW_{km}\right|\Big/\sum_{k=1}^{N_i}\left|IW_{km}\right|\right)\times\left|LW_{mn}\right|\right)\right\}}\times 100 \qquad (4)$$

where RI_j is the relative importance of the jth input variable on the output variable, N_i and N_h are the number of input variables and neurons in hidden layer, respectively (N_i =4, N_h =4 in this study), IW and LW are the connection weights, and subscript "n" refer to output response (n=1) [31].

The relative importance of electrospinning parameters on the value of CA calculated by Equation (4) and is shown in Figure 6. It can be seen that, all of the input variables have considerable effects on the CA of electrospun fiber mat. Nevertheless, the solution concentration with relative importance of 49.69% is found to be most important factor affecting the CA of electrospun nanofibers. These results are in close agreement with those obtained with RSM.

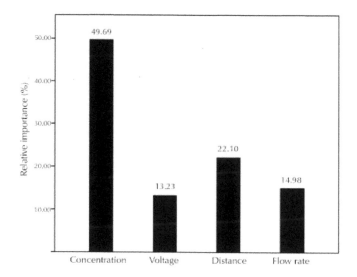

FIGURE 6   Relative importance of electrospinning parameters on the CA of electrospun fiber mat.

10.3.4   OPTIMIZING THE CA OF ELECTROSPUN FIBER MAT

The optimal values of the electrospinning parameters were established from the quadratic form of the RSM. Independent variables (solution concentration, applied voltage, spinning distance, and volume flow rate) were set in range and dependent variable (CA) was fixed at minimum. The optimal conditions in the tested range for minimum CA of electrospun fiber mat are shown in Table 5. This optimum condition was a predicted value, thus to confirm the predictive ability of the RSM model for response, a further electrospinning and CA measurement was carried out according to the optimized conditions and the agreement between predicted and measured responses was verified. Figure 7 shows the SEM, average fiber diameter distribution and corresponding CA image of electrospun fiber mat prepared at optimized conditions.

TABLE 5   Optimum values of the process parameters for minimum CA of electrospun fiber mat

Solution concentration (wt.%)	Applied voltage (kV)	Spinning distance (cm)	Volume flow rate (ml/h)	Predicted CA (°)	Observed CA (°)
13.2	16.5	10.6	2.5	20	21

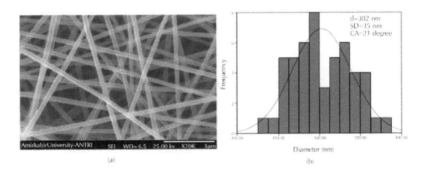

(a)   (b)

(c)

FIGURE 7   (a) SEM image, (b) fiber diameter distribution, and (c) CA of electrospun fiber mat prepared at optimized conditions.

10.3.5   COMPARISON BETWEEN RSM AND ANN MODEL

Table 6 gives the experimental and predicted values for the CA of electrospun fiber mat obtained from RSM as well as ANN model. It is demonstrated that both models performed well and a good determination coefficient was obtained for both RSM and ANN. However, the ANN model shows higher determination coefficient ($R^2 = 0.965$) than the RSM model ($R^2 = 0.958$). Moreover, the absolute percentage error in the ANN prediction of CA was found to be around 5.94%, while for the RSM model, it was around 7.83%. Therefore, it can be suggested that the ANN model shows more accurately result than the RSM model. The plot of actual and predicted CA of electrospun fiber mat for RSM and ANN is shown in Figure 8.

TABLE 6   Experimental and predicted values by RSM and ANN models

No.	Experimental	Predicted		Absolute error (%)	
		RSM	ANN	**RSM**	**ANN**
1	44	47	48	6.41	9.97
2	54	54	54	0.78	0.46
3	61	59	61	3.70	0.42
4	65	66	61	2.06	6.06
5	38	39	38	2.37	0.54
6	49	47	49	5.10	0.68
7	51	51	51	0.35	0.45
8	56	58	56	4.32	0.17
9	48	45	60	6.17	24.37
10	30	31	27	4.93	9.35
11	35	36	31	2.34	11.15
12	22	22	21	1.18	4.15
13	30	30	32	1.33	6.04
14	33	28	33	13.94	0.60
15	25	24	25	5.04	0.87
16	26	26	26	0.27	1.33
17	29	26	26	10.10	9.16
18	28	26	26	6.89	5.91
19	25	26	26	4.28	5.38
20	24	26	26	8.63	9.77
21	21	26	26	24.14	25.45
22	31	30	31	2.26	0.57
23	35	31	35	10.34	0.66
24	33	36	32	8.18	2.18
25	37	37	37	0.59	0.34
26	19	29	21	52.11	10.23
27	28	30	30	7.07	8.20
28	39	34	31	12.05	21.30

TABLE 6   *(Continued)*

29	36	35	36	1.72	0.04
30	20	25	20	26.30	2.27
R^2		0.958	0.965		
Mean absolute error (%)				7.83	5.94

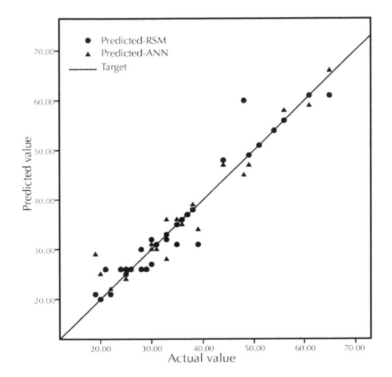

FIGURE 8   Comparison between the actual and predicted contact angle of electrospun nanofiber for RSM and ANN model.

10.4   CONCLUSION

The morphology and properties of electrospun nanofibers depends on many processing parameters. In this work, the effects of four electrospinning parameters namely; solution concentration (wt.%), applied voltage (kV), tip to collector distance (cm), and volume flow rate (ml/h) on CA of PAN nanofiber mat were investigated using two different quantitative models, comprising RSM and ANN. The RSM model confirmed that solution concentration was the most significant parameter in the CA of electrospun fiber mat. Comparison of predicted CA using RSM and ANN were also studied. The obtained results indicated that both RSM and ANN model shows a very good relationship between the experimental and predicted CA values. The ANN model shows higher determination coefficient ($R^2=0.965$) than the RSM model. Moreover, the absolute percentage error of prediction for the ANN model was much lower than that for RSM model, indicating that ANN model had higher modeling performance than RSM model. The minimum CA of electrospun fiber mat estimated by RSM equation obtained at conditions of 13.2 wt.% solution concentration, 16.5 kV of the applied voltage, 10.6 cm of tip to collector distance, and 2.5 ml/h of volume flow rate.

KEYWORDS

- **Fiber Spinning Systems**
- **Artificial Neural Network**
- **Analysis of Variance**

REFERENCES

1. Miwa, M., Nakajima, A., Fujishima, A., Hashimoto, K., and Watanabe, T. Langmuir, 16, 5754 (2000).
2. Öner, D. and McCarthy, T. J. Langmuir, 16, 7777 (2000).
3. Abdelsalam, M. E., Bartlett, P. N., Kelf, T., and Baumberg, J. Langmuir, 21, 1753 (2005).
4. Nakajima, A., Hashimoto, K., Watanabe, T., Takai, K., Yamauchi, G., and Fujishima, A. Langmuir, 16, 7044 (2000).
5. Zhong, W., Liu, S., Chen, X., Wang, Y., and Yang, W. Macromolecules, 39, 3224 (2006).
6. Shams Nateri, A. and Hasanzadeh, M. J. Comput. Theor. Nanosci., 6, 1542 (2009).
7. Kilic, A., Oruc, F., and Demir, A. Text. Res. J., 78, 532 (2008).
8. Reneker, D. H. and Chun, I. Nanotechnology, 7, 216 (1996).
9. Shin, Y. M., Hohman, M. M., Brenner, M. P., and Rutledge, G. C. Polymer, 42, 9955 (2001).
10. Reneker, D. H., Yarin, A. L., Fong, H., and Koombhongse, S. J. Appl. Phys., 87, 4531 (2000).
11. Zhang, S., Shim, W. S., and Kim, J. Mater. Design, 30, 3659 (2009).
12. Yördem, O. S., Papila, M., and Menceloğlu, Y. Z. Mater. Design, 29, 34 (2008).
13. Chronakis, I. S. J. Mater. Process. Tech., 167, 283 (2005).
14. Dotti, F., Varesano, A., Montarsolo, A., Aluigi, A., Tonin, C., and Mazzuchetti, G. J. Ind. Text., 37, 151 (2007).
15. Lu, Y., Jiang, H., Tu, K., and Wang, L. Acta Biomater., 5, 1562 (2009).
16. Lu, H., Chen, W., Xing, Y., Ying, D., and Jiang, B. J. Bioact. Compat. Pol., 24, 158 (2009).
17. Nisbet, D. R., Forsythe, J. S., Shen, W., Finkelstein, D. I., and Horne, M. K. J. Biomater. Appl., 24, 7 (2009).
18. Ma, Z., Kotaki, M., Inai, R., and Ramakrishna, S. Tissue Eng., 11, 101 (2005).
19. Hong, K. H. Polym. Eng. Sci., 47, 43 (2007).
20. Zhang, W. and Pintauro, P. N. ChemSusChem., 4, 1753 (2011).
21. Lee, S. and Obendorf, S. K. Text. Res. J., 77, 696 (2007).

22. Myers, R. H., Montgomery, D. C., and Anderson-cook, C. M. Response surface methodology: process and product optimization using designed experiments, 3rd ed., John Wiley and Sons, USA (2009).
23. Gu, S. Y., Ren, J., and Vancso, G. J. Eur. Polym. J., 41, 2559 (2005).
24. Dev, V. R. G., Venugopal, J. R., Senthilkumar, M., Gupta, D., and Ramakrishna, S. J. Appl. Polym. Sci., 113, 3397 (2009).
25. Galushkin, A. L. Neural networks Theory, Springer, Moscow Institute of Physics & Technology (2007).
26. Ma, M., Mao, Y., Gupta, M., Gleason, K. K., and Rutledge, G. C. Macromolecules, 38, 9742 (2005).
27. Haghi, A. K. and Akbari, M. Phys. Status. Solidi. A., 204, 1830 (2007).
28. Ziabari, M., Mottaghitalab, V., and Haghi, A. K. in Nanofibers: Fabrication, Performance, and Applications. W. N. Chang (Ed.), Nova Science Publishers, USA (2009).
29. Ramakrishna, S., Fujihara, K., Teo, W. E., Lim, T. C., and Ma, Z. An Introduction to Electrospinning and Nanofibers, World Scientific Publishing, Singapore (2005).
30. Zhang, S., Shim, W. S., and Kim, J. Mater. Design, 30, 3659 (2009).
31. Kasiri, M. B., Aleboyeh, H., and Aleboyeh, A. Environ. Sci. Technol., 42, 7970 (2008).

CHAPTER 11

THE INFLUENCE OF CRYSTALLINE MORPHOLOGY ON FRACTAL SPACE FORMATION FOR NANOCOMPOSITES POLYMER/ORGANOCLAY

K. S. DIBIROVA, G. V. KOZLOV, G. M. MAGOMEDOV, and G. E. ZAIKOV

CONTENTS

11.1   INTRODUCTION

It has been shown that crystalline phase morphology in nanocomposites polymer/organo-clay with semicrystalline matrix defines the dimension of fractal space, in which the indicated nanocomposites structure is formed. In its turn, this dimension influences strongly on both deformational behavior and mechanical characteristics of nanocomposites.

It has been shown earlier [1, 2], that particles (aggregates of particles) of filler (nanofiller) form network in polymer matrix, possessing fractal (in the general case – multifractal) properties and characterized by fractal (Hausdorff) dimension D_n. Hence, polymer matrix structure formation in nanocomposites can be described not in Euclidean space, but in fractal one. This circumstance tells to a considerable degree on both structure and properties of nanocomposites. As it has been shown in work [2], polymer nanocomposites properties change is defined by polymer matrix structure change, which is due to nanofiller introduction. So, the authors [3] demonstrated that the introduction of organoclay in high density polyethylene (HDPE) resulted in matrix polymer crystalline morphology change, that is, in spherolites size increasing about twice. Therefore the present work purpose is the study of polymer matrix crystalline morphology influence on structure and properties of nanocomposites high density polyethylene/Na+-montmorillonite (HDPE/MMT). This study was performed within the framework of fractal analysis [4].

11.2   EXPERIMENTAL

As a matrix polymer HDPE with melt flow index of ~ 1.0 g/10 min and crystallinity degree K of 0.72, determined by samples density, manufactured by firm Huntsman LLC, was used. As nanofiller organoclay Na+-montmorillonite of industrial production of mark Cloisite 15, supplied by firm Southern Clay (USA), was used. A maleine anhydride (MA) was applied as a coupling agent. Conventional signs and composition of nanocomposites HDPE/MMT are listed in Table 1 [3].

TABLE 1   Composition and spherolites average diameter of nanocomposites HDPE/MMT

Sample conventional sign	MA contents, mass. %	MMT contents, mass. %	Spherolites average diameter D_{sp}, mcm
A	-	-	5.70
B	1.0	-	5.80
C	-	1.0	11.62
D	-	2.5	10.68
E	-	5.0	10.75
F	1.0	1.0	11.68
G	2.5	2.5	11.12
H	5.0	5.0	11.0
I	5.0	2.5	11.30

Compositions HDPE/MA/MMT were prepared by components mixing on twin screw extruder of mark Haake TW 100 at temperatures 390–410 K. Samples with thickness of 1 mm were obtained on one-screw extruder Killion [3].

Uniaxial tension mechanical tests have been performed on apparatus Rheometric Scientific Instrument (RSA III) according to ASTM D882-02 at temperature 293 and strain rate $2 \cdot 10^{-3}$ s^{-1}. The morphology of nanocomposites HDPE/MMT was studied with the help of polarized optical microscope Zeiss with magnification 40 and thus obtained spherolites average diameter D_{sp} in this way is also adduced in Table 1.

11.3 DISCUSSION AND RESULTS

As it is known [5], the fractal dimension of an object is a function of space dimension, in which it is formed. In the computer model experiment this situation is considered as fractals behavior on fractal (but not Euclidean) lattices [6]. The space (or fractal lattice) dimension D_n can be determined with the aid of the following equation [1]:

$$v_F = \frac{2.5}{2 + D_n},$$

(1)

where v_F is Flory exponent, connected with macromolecular coil dimension D_f by the following relationship [1]:

$$v_F = \frac{1}{D_f}.$$

(2)

In its turn, the dimension D_f value for linear polymers the following simple equation gives [7]:

$$D_f = \frac{2d_f}{3},$$

(3)

where d_f is the fractal dimension of polymeric material structure, determined as follows [8]:

$$d_f = (d - 1)(1 + v),$$

(4)

where d is the dimension of Euclidean space, in which a fractal is considered (it is obvious, that in our case $d=3$), v is Poisson's ratio, estimated according to the results of mechanical tests with the help of the relationship [9]:

$$\frac{\sigma_Y}{E} = \frac{1-2v}{6(1+v)},$$

(5)

where σ_Y is yield stress, E is elasticity modulus.

In Figure 1 the dependence of fractal space (fractal lattice) dimension D_n, in which the studied nanocomposites structure is formed, on spherolites mean diameter D_{sp} is adduced. As one can see, the correlation $D_n(D_{sp})$ is linear and described analytically by the following empirical equation:

$$D_n = 2.1 \times 10^{-2} D_{sp} + 2.5,$$

(6)

where value D_{sp} is given in mcm.

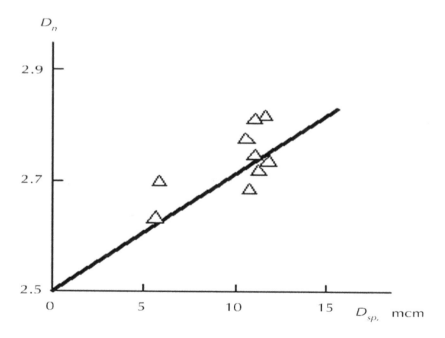

FIGURE 1   The dependence of space dimension D_n, in which nanocomposite structure is formed, on spherolites average diameter D_{sp} for nanocomposites HDPE/MMT.

From the equation (6) it follows, that the minimum value D_n is achieved at $D_{sp}=0$ and is equal to 2.5, that according to the equations (1)-(3) corresponds to structure fractal dimension d_f 2.17. Since the greatest value of any fractal dimension for real objects, including D_n at that, cannot exceed 2.95 [8], then from the equation (6) the limiting value D_{sp} for the indicated matrix polymer can be evaluated, which is equal to

~ 21.5 mcm. Let us also note that the large scatter of the data in Figure 1 is due to the difficulty of the value D_{sp} precise determination.

As it is known [7], in semicrystalline polymers deformation process partial melting-recrystallization (mechanical disordering) of a crystalline phase can be realized, which is described quantitatively within the framework of a plasticity fractal theory [10]. According to the indicated theory Poisson's ratio value v_Y at a yield point can be evaluated as follows:

$$v_Y = \chi v + 0.5(1-\chi)\text{,} \tag{7}$$

where χ is a relative fraction of elastically deformed polymer, v is Poisson's ration in elastic strains region, determined according to the equation (5), and v_Y value is accepted equal to 0.45 [7].

The calculation of a relative fraction of crystalline phase χ_{cr}, subjecting to mechanical disordering, can be performed according to the equation [7]:

$$\chi_{cr} = \chi - \alpha_{am} - \phi_{cl} \tag{8}$$

where α_{am} is an amorphous phase relative fraction, which is equal to $(1-K)$, φ_{cl} is a relative fraction of local order domains (nanoclusters), which can be determined with the aid of the following fractal relationship [7]:

$$d_f = 3 - 6 \times 10^{-10} \left(\frac{\phi_{cl}}{SC_\infty} \right)^{1/2}, \tag{9}$$

where S is cross-sectional area of macromolecule, which is equal to 14.4 Å^2 for HDPE [7], C is characteristic ratio, which is an indicator of polymer chain statistical flexibility [11], and connected with the dimension d_f by the following relationship [7]:

$$C_\infty = \frac{2d_f}{d(d-1)(d-d_f)} + \frac{4}{3}. \tag{10}$$

As it is known [7], parameter χ_{cr} effects essentially on deformational behavior and mechanical properties of semicrystalline polymers. In Figure. 2 the dependence $\chi_{cr}(D_n)$ is adduced for the studied nanocomposites, which turns out to be linear, that allows to describe it analytically as follows:

$$\chi_{cr} = 1.88(D_n - 2.55). \tag{11}$$

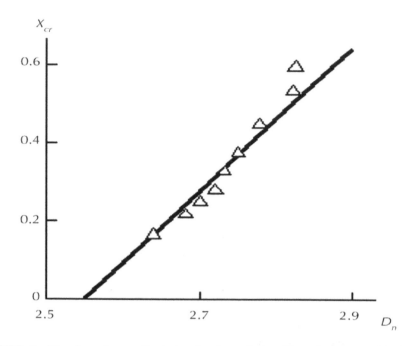

FIGURE 2 The dependence of relative fraction of crystalline phase χ_{cr}, subjecting to mechanical disordering, on space dimension D_n for nanocomposites HDPE/MMT.

From the equation (8) it follows that the greatest value χ_{cr} (χ_{cr}^{max}) is achieved at the following conditions: $\chi=1.0$ and $\varphi_{cl}=0$. In this case the condition $\chi_{cr}^{max} = K$ is realized, that was to be expected from the most common considerations. For the studied nanocomposites the value $\chi_{cr}^{max}=K=0.72$ is achieved according to the equation (11) at D_n 2.933, that is close to the indicated above limiting value $D_n=2.95$ [8]. The minimum value $\chi_{cr}=0$ according to the equation (11) is achieved at $D_n=2.55$ or, according to the formulas (1)-(3), at $d_f=2.73$. As it is known [7], the value d_f can be determined alternatively as follows:

$$d_f \approx 2 + K \cdot \qquad (12)$$

From the equations (11) and (12) it follows, that a common variation $\chi_{cr}=0-0.72$ is realized at the constant value $K=0.72$, that is, this parameter does not depend on crystallinity degree and it is defined only by polymer matrix crystalline morphology change.

As it is known [4], the parameter χ_{cr} influences essentially on nanocomposites polymer/organoclay properties. One from the most important mechanical characteristics of polymeric materials, namely, elasticity modulus E depends on the value χ_{cr} as follows:

$$E = \left(40 + 54.9 \chi_{cr}\right) \sigma_Y. \tag{13}$$

In Figure 3 the comparison of experimental E and calculated according to the equation (13) E^T elasticity modulus values for the considered nanocomposites is adduced. In this case the value χ_{cr} was determined according to the relationship (11). As it follows from the data of Figure 3, a theory and experiment good correspondence is obtained (the average discrepancy between E^T and E makes up ~ 14 %), that will be enough for preliminary estimations performance.

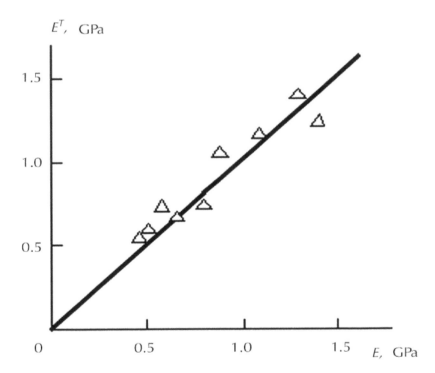

FIGURE 3   The relation between experimental E and calculated according to the equations (11) and (13) E^T elasticity modulus values for nanocomposites HDPE/MMT.

11.4   CONCLUSION

Thus, the present work results have shown that the fractal dimension of space, in which nanocomposites structure is formed, is defined by their crystalline polymer phase morphology and does not depend on crystallinity degree. The indicated dimension defines unequivocally partial melting-recrystallization process at nanocomposites deformation and influences strongly on their mechanical characteristics.

KEYWORDS

- **Elasticity Modulus**
- **Fractal Space**
- **Morphology**
- **Nanocomposite**
- **Organoclay**
- **Semicrystalline Polymer**

REFERENCES

1. Kozlov, G. V., Yanovskii, Yu.G., and Zaikov, G. E. *Structure and Properties of Particulate-Filled Polymer Composites: the Fractal Analysis.* Nova Science Publishers, Inc., New York, p. 282 (2010).
2. Miktaev, A. K., Kozlov, G. V., and Zaikov, G. E. *Polymer Nanocomposites: Variety of Structural Forms and Applications.* Nova Science Publishers, Inc., New York, p. 319 (2008).
3. Ranade, A. Nayak, K, Fairbrother, D., and D'Souza, N. A. *Polymer,* **46**(23), 7323–7333 (2005).
4. Kozlov, G. V. and Miktaev, A. K. *Structure and Properties of Nanocomposites Polymer/organoclay.* LAP LAMBERT Academic Publishing GmbH, Saarbrücken, p. 318 (2013).
5. Aharony, A. and Harris, A. B. *J. Stat. Phys.,* **54**(3/4), 1091–1097 (1989).
6. Vannimenus, J. *Physica D,* **38**(2), 351–355 (1989).
7. Kozlov, G. V. and Zaikov, G. E. *Structure of the Polymer Amorphous State.* Brill Academic Publishers, Utrecht, Boston, p. 465 (2004).
8. Balankin, A. S. *Synergetics of Deformable Body.* Publishers of Ministry Defence of SSSR, Moscow, p. 404 (1991).
9. Kozlov, G. V. and Sanditov, D. S. *Anharmonic Effects and Physical-Mechanical Properties of Polymers.* Nauka, Novosibirsk, p. 261 (1994).
10. Balankin, A. S. and Bugrimov, A. L. *Vysokomolek. Soed. A,* **34**(3), 129–132 (1992).
11. Budtov, V. P. Physical *Chemistry of Polymer Solutions.* Khimiya, Sankt-Peterburg, p. 384 (1992).

CHAPTER 12

THE FRACTAL MODEL OF COKE RESIDUE FORMATION FOR COMPOSITES HIGH DENSITY POLYETHYLENE/ALUMINIUM HYDROXIDE

I. V. DOLBIN, G. V. KOZLOV, G. E. ZAIKOV, and A. K. MIKITAEV

CONTENTS

12.1   INTRODUCTION

The structural (fractal) analysis of coke residue formation at composites high density polyethylene/aluminum hydroxide combustion was performed. It has been shown that aluminum hydroxide particles aggregates formation results in such situation, when the indicated particles decomposition is realized in loose surface layers of aggregates, whereas densely-packed central regions form coke residue.

At present many methods of polymeric materials flame-resistance enhancement with the aid of special additions, which are known as antipyrene, are being developed. They act on different ways of combustion origin and spreading: on material itself, its combustion heat and air flow. Metal hydroxides can be used as such effective antipyrens [1].

The authors of the work [2] used aluminum hydroxide Al(OH)$_3$ for flame-resistance enhancement of high density polyethylene (HDPE). The coke residue value W_{cr} determination according to the data of thermal analysis has shown [2], that hydroxide of aluminum introduction in HDPE contributes to W_{cr} increasing in comparison with the initial polymer, for which $W_{cr}=0$, i.e. it enhances HDPE flame-resistance. The present work purpose is the structural treatment of coke residue formation in composites HDPE/Al(OH)$_3$ receiving. This treatment will be obtained within the framework of fractal analysis [3, 4].

12.2   EXPERIMENTAL

Gas-phase HDPE of industrial production of mark HDPE-273, GOST 16338-85 with average molecular weight of 2.5×10^5 and crystallinity degree 0.69 was used as a matrix polymer. Al(OH)$_3$ of mark ALFRIMAL 106, having density of 2400 kg/m^3 and specific surface by BET 30 m^2/g was used as a filler (antipyrene). Al(OH)$_3$ contents in composites HDPE/Al(OH)$_3$ changed within the limits of 5-50 mass %.

Compositions HDPE/Al(OH)$_3$ were prepared by components mixing in melt on twin screw extruder Thermo Haake, model Reomex RTW 25/42, production of German Federal Republic. Mixing was performed at temperatures 463-503 K and screw speed of 50 rpm during 5 min. Testing samples were obtained by casting under pressure method on a casting machine Test Sample Molding Apparate RR/TS of firm Ray-Ran (Taiwan) at temperature 483 K and pressure 43 MPa.

The value of coke residue measured in thermal tests on derivatograph MOM Q 1500 (Hungary), was used for compositions HDPE/Al(OH)$_3$ flame-resistance estimation.

12.3   DISCUSSION AND RESULTS

The relation of experimental values of the coke residue W_{cr} and antipyrene mass contents W_n for the considered composites is adduced in Figure 1. As one can see, at small W_n of the order of 5-10 mass % the indicated parameters approximate equality is observed, but at $W_n > 10$ mass % W_{cr} value decreases systematically in comparison with W_n. The aggregation of Al(OH)$_3$ particles at their contents increasing can be one of the possible reasons of the observed effect. Let us consider this question in more detail.

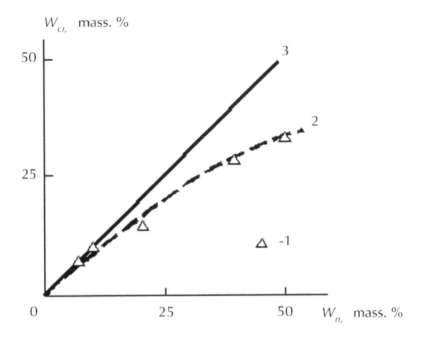

FIGURE 1   The relation between coke residue value W_{cr} and antipyrene mass contents W_n for composites HDPE/Al(OH)$_3$. 1 – experimental data; 2 – calculation according to the equation (7); 3 – the relation 1:1.

The filler aggregates diameter D_{agr} can be evaluated according to the following empirical equation [3]:

$$D_{agr} = 5.24\,\phi_n^{1/2}\,,$$ (1)

where ϕ_n is Al(OH)$_3$ volume content, which is determined according to the well-known equation [3]:

$$\phi_n = \frac{W_n}{\rho_n}\,,$$ (2)

where Al(OH)$_3$ particles aggregates density ρ_n is accepted equal to 2000 kg/m^3, i.e. somewhat smaller than compact material density ρ_{comp} (ρ_{comp}=2400 kg/m^3).
    The estimations according to the equation (1) have shown that D_{agr} value for Al(OH)$_3$ aggregates varies within the limits of 0.83-2.62 mcm. Using the calibrating plots, adduced in work [4], Al(OH)$_3$ aggregates surface fractal dimension d_{surf} value can be estimated as equal to ~ 2.20 and their structure fractal dimension d_f value as

equal to ~ 2.35. Further the indicated aggregates specific surface S_u can be determined, using the following formula [4]:

$$S_u = 410 \left(\frac{D_{agr}}{2} \right)^{d_{surf} - d}, \text{ m2/g,} \qquad (3)$$

where D_{agr} is given in nm, d is dimension of Euclidean space, in which a fractal is considered (it is obvious, that in our case $d=3$).

Al(OH)$_3$ fractal aggregates density ρ_{fr} can be calculated according to the equation [5]:

$$\rho_{fr} = \frac{6}{S_u D_{agr}} \cdot \qquad (4)$$

The estimations according to the equation (4) have shown ρ_{fr} reduction from 2190 up to 1730 kg/m^3 within the range of W_n=5–50 mass %, that corresponds well to the accepted above approximation of ρ_n=2000 kg/m^3.

As it is known [4], fractal clusters in their structure center form densely-packed region of size a, the value of which can be determined with the aid of the following equation [4]:

$$\rho_{fr} = \rho_{comp} \left(\frac{D_{agr}}{a} \right)^{d_f - d} \cdot \qquad (5)$$

It should be assumed that densely-packed central region of Al(OH)$_3$ fractal aggregates does not let the air out and, hence, is not subjected to decomposition. The indicated circumstance defines the observed in Figure 1 parameters W_{cr} and W_n discrepancy at the condition $W_{cr} < W_n$. The relative fraction φ_f of aggregate of Al(OH)$_3$ friable material, subjecting to decomposition, can be estimated as follows:

$$\varphi_f = \frac{D_{agr}^3 - a^3}{D_{agr}^3} \cdot \qquad (6)$$

Then antipyrene Al(OH)$_3$ fraction, which does not subject to decomposition, i.e. coke residue fraction W_{cr}, is determined according to the equation:

$$W_{cr} = W_n \left(1 - \phi_f \right). \qquad (7)$$

In Figure 1 theoretical correlation $W_{cr}(W_n)$, calculated according to the equation (7), is shown by a stroken line. As one can see, it corresponds well to the experimental data, that confirms correctness of the proposed above fractal (structural) model of coke residue formation for composites HDPE/Al(OH)$_3$.

Let us consider further the correctness of Al(OH)$_3$ aggregates diameter estimation according to the empirical relationship (1). Within the framework of an irreversible aggregation model at aggregates origination sites ("seeds") large number availability, which corresponds to the considered case, these aggregates critical radius R_c, at which their growth ceases, can be estimated with the aid of the relationship [6]:

$$c \sim R_c^{d_f} , \tag{8}$$

where c is the initial concentration of aggregating particles.

In the considered case $c = \varphi_n$, $2R_c = D_{agr}^T$ and the value d_f is accepted, as earlier, equal to 2.35. In Figure 2 the comparison of Al(OH)$_3$ aggregates diameter D_{agr}^T theoretical values, calculated by the indicated mode, and the values D_{agr}, calculated according to the equation (1), is adduced. As it follows from this comparison, the relation between D_{agr}^T and D_{agr} is approximated well by the linear correlation, passing through coordinate's origin, and is described analytically by the following empirical equation:

$$D_{agr} = 4.72\varphi_n^{1/d_f} , \text{ mcm.} \tag{9}$$

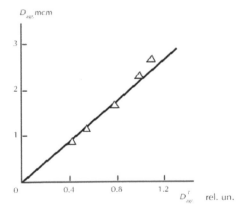

FIGURE 2   The comparison of values of fractal aggregates diameter D_{agr} (the equation (1)) and D_{agr}^T (the relationship (8)) for composites HDPE/Al(OH)$_3$.

As it is known [7], when fractal aggregate grows at the expence of diffusible particles irreversible addition, then new particles form only a small part of a total number N of aggregating particles. These new particles present themselves a growing interfacial boundary of the fractal cluster, consisting of N_i particles. For three-dimensional Euclidean space the authors [7] received the following relationship between N_i and N:

$$N_i \sim N^{0.7}. \tag{10}$$

One can be assume, that precisely N_i particles, making up just formed and, consequently, the most friable part of fractal aggregate, are subjected to decomposition in combustion process of composites HDPE/Al(OH)$_3$. The particles number N in the indicated aggregate can be estimated with the aid of the following equation [4]:

$$\frac{D_{agr}}{2} = \left(\frac{NS_p}{\pi\eta}\right)^{1/2}, \tag{11}$$

where S_p is cross-sectional area of particles, forming a cluster, η is density of particles packing in the cluster, which is equal to 0.74 [4].

The number of particles $N_d = N_i$, subjecting to decomposition in composites HDPE/Al(OH)$_3$ combustion process, can be determined as follows:

$$N_d = N\varphi_f. \tag{12}$$

In Figure 3 the dependence $N_d(N)$ in double logarithmic coordinates for composites HDPE/Al(OH)$_3$ is adduced, which turns out to be linear, that allows to determine the exponent in the relationship $N_d(N)$, which is equal to ~ 0.78. The indicated exponent is close by absolute value to the corresponding exponent in the relationship (10). This correspondence assumes, that Al(OH)$_3$ particles, forming an external layer (interfacial boundary) of the indicated particles fractal aggregates, are subjected to decomposition in composites HDPE/Al(OH)$_3$ combustion process, that is due to this layer loosely-packed structure.

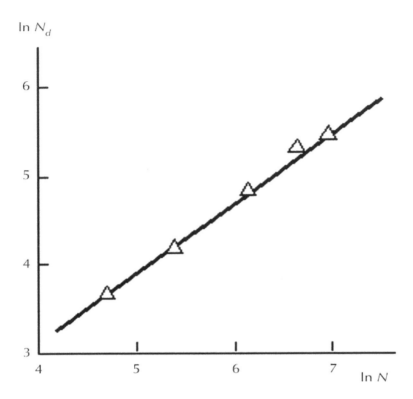

FIGURE 3 The relation between antipyrene $Al(OH)_3$ particles number N_d, subjecting to decomposition, and particles total number N in fractal aggregate in double logarithmic coordinates for composites $HDPE/Al(OH)_3$.

12.4 CONCLUSION

Thus, the proposed in the present work structural (fractal) treatment explains quantitatively the coke residue formation process at composites $HDPE/Al(OH)_3$ combustion. The antipyrene (in the considered case - $Al(OH)_3$) decomposition is realized in the surface (interfacial) layers of fractal aggregates, formed by $Al(OH)_3$ particles in their aggregation process. This process in the given case is due to the indicated layers friable structure that allows an easy access to them the air, necessary for decomposition.

KEYWORDS

- **Composite**
- **Polyethylene**
- **Aluminum Hydroxide**
- **Combustion**
- **Coke Residue**
- **Aggregation**

REFERENCES

1. Aseeva, R. M. and Zaikov, G. E. *Combustion of Polymeric Materials*, Nauka, Moscow, p. 278 (1981).
2. Borukaev, T. A. and Khatsukova, R. B. Mater. of VI International Sci.-Pract. Conf. "New Polymeric Composite Materials", *KBSU*, Nal'chik, 126–131 (2010).
3. Kozlov, G. V., Yanovskii, Yu. G., and Zaikov, G. E. *Structure and Properties of Particulate-Filled Polymer Composites: the Fractal Analysis*, Nova Science Publishers, Inc., New York, p. 282 (2010).
4. Mikitaev, A. K., Kozlov, G. V., and Zaikov, G. E. *Polymer Nanocomposites: Variety of Structural Forms and Applications*, Nova Science Publishers, Inc., New York, p. 319 (2008).
5. Bobryshev, A. N., Kozomazov, V. N., Babin, L. O., and Solomatov, V. I. *Synergetics of Composite Materials*, NPO ORIUS, Lipetsk, p. 154 (1994).
6. Witten, T. A. and Meakin, P. *Phys. Rev. B*, **28**(10), 5632–5642 (1983).
7. Meakin, P. and Witten, T. A. *Phys. Rev. A*, **28**(5), 2985–2988 (1983).

CHAPTER 13

THE STRUCTURAL MODEL OF NANOCOMPOSITES POLY(VINYL CHLORIDE)/ORGANOCLAY FLAME-RESISTANCE

I. V. DOLBIN, G. V. KOZLOV, G. E. ZAIKOV, and A. K. MIKITAEV

CONTENTS

13.1   INTRODUCTION

It has been shown within the framework of strange (anomalous) diffusion conception that instantaneous jumps ("Levy's flights") of combustion front from one region of polymeric material into another increase sharply this material flammability. Distance between nanofiller particles decreasing reduces such jumps intensity, increasing thereby material flame-resistance. The fractal time of combustion enhancement results in "Levy's flights" intensification and vice versa.

At it is known [1, 2], organoclay introduction in polymer increases essentially its flame-resistance. This effect usually is explained by "barrier effect" appearance that is organoclay nanoparticles form a kind of barriers, preventing combustion front propagation. In this process structural features of organoclay influence essentially on flame-resistance of nanocomposites, filled with it. For example, organoclay content increasing results in flame-resistance enhancement and exfoliated organoclay suppresses the ability to combustion more effectively that intercalated one [1, 2]. The present work purpose is structural analysis of nanocomposites polymer/organoclay flame-resistance enhancement within the framework of fractal analysis.

13.2   EXPERIMENTAL

The poly(vinyl chloride) plasticate (PVC) of mark U40-13A of standard recipe 8/2, prepared according to GOST 5962-72, was used. The montmorillonite (MMT), prepared from natural clay according to the technique [3], with the cation exchange capacity of 95 mg–equivalent/100 g of clay was used [4].

Nanocomposites PVC/MMT with organoclay contents of 1–7 mass% were obtained by blending in twin speed blender R600/HC 2500 production of firm "Diosna", the design of which ensures turbulent blending with nanocomposition homogenization high extent and blowing off with hot air. After plasticate of PVC intensive intermixing with organoclay in a blender at temperature 383–393 K up to obtaining high-disperse free-flowing mixture the composition was cooled up to temperature 313 K and then it was processed on twin screw extruder Thermo Haake, model Reomex RTW 25/42, production of German Federal Republic, at temperatures 398–423 K and screw speed of 48 rpm. Testing samples were obtained by casting under pressure method of granulated nanocomposites on a casting machine Test Sample Molding Apparate RR/TS MP of firm Ray-Ran (Taiwan) at temperature 443 K and pressure 12 MPa during 4 min [4].

The flame-resistance (putting out a fire time) is measured on device UL-94 of firm Noselab (Italy) according to GOST 28157-80 [4].

Uniaxial tension mechanical tests have been performed on the samples in the shape of two-sided spade with sizes according to GOST 11262-80. The tests have been conducted on universal testing apparatus Gotech Testing Machine CT-TCS 2000, production of German Federal Republic, at temperature 293 K and strain rate $\sim 2 \square 10^{-3}$ s^{-1} [4].

13.3   DISCUSSION AND RESULTS

The authors [5] formulated the fractional equation of transport processes, having the following form:

$$\frac{\partial^{\alpha}\psi}{\partial t^{\alpha}} = \frac{\partial^{2\beta}}{\partial r^{2\beta}}(B\psi), \tag{1}$$

where, $\psi = \psi(t, r)$ is particles distribution function, $\partial^{2\beta}/\partial r^{2\beta}$ is Laplacian in d-dimensional Euclidean space, representing the ratio on generalized transport coefficient and d. The introduction of fractional derivatives $\partial^{\alpha}/\partial t^{\alpha}$ and $\partial^{2\beta}/\partial r^{2\beta}$ allows to take into account memory effects (α) and nonlocality effects (β) in the context of a single mathematical formalism [5].

The introduction of fractional derivative $\partial^{\alpha}/\partial t^{\alpha}$ in the kinetic Equation (1) also allows taking into account random walks in fractal time (RWFT)—a "temporal component" of strange dynamic processes in turbulent mediums [5]. The absence of some appreciable jumps in particles behavior serves as a distinctive feature of RWFT; in addition mean-square displacement grows with t as t^{α}. Parameter α has the sense of fractal dimension of "active" time, in which real particles walks look as random process: active time interval is proportional to t^{α} [5].

In its turn, the exponent 2β in the Equation (1) takes into account particles instantaneous jumps ("Levy's flights") from one region into another. Thus, the exponents ratio α/β gives the ratio of RWFT and "Levy's flights" contact frequencies. The value α/β is equal to [5]:

$$\frac{\alpha}{\beta} = \frac{d_s}{d}, \tag{2}$$

where d_s is spectral dimension of polymeric material structure, d is dimension of Euclidean space, in which a fractal is considered (it is obvious, that in our case $d = 3$).

The structure fractal (Hausdorff) dimension d_f can be determined as follows [6]:

$$d_f = (d-1)(1+v), \tag{3}$$

where, v is Poisson's ratio, estimated by mechanical tests results with the aid of the Equation (6):

$$\frac{\sigma_Y}{E} = \frac{1-2v}{6(1+v)}, \tag{4}$$

where σ_Y is yield stress, E is elasticity modulus.

For linear polymers a macromolecular coil fractal dimension D_f is calculated according to the Equation (6):

$$D_f = \frac{2d_f}{3}. \tag{5}$$

Further the value d_s can be determined with the aid of the relationship (6):

$$D_f = \frac{d_s(d+2)}{2}.$$

(6)

The parameter α value is calculated as follows [7]:

$$\alpha = 0.5(2 - D_f).$$

(7)

In case of polymeric material sample combustion this reaction completeness degree Q can be determined as follows. As it is known [7], diffusion process duration τ is calculated according to the equation:

$$\tau = \frac{l_{com}^2}{6D},$$

(8)

where, l_{com} is polymeric material burning depth, D is diffusivity of this material. The value Q is determined as the ratio:

$$Q = \frac{2l_{com}}{l},$$

(9)

where, l is polymer sample initial thickness, which is equal to 4 mm in the considered case.

Diffusivity coefficient D_m for PVC matrix plasticate can be determined from the Equation (8) in supposition, that combustion front achieves the sample middle (l_{com} = 2 mm) and its achievement time τ is equal to experimental value of putting out a fire time τ_p = 4.5 s. Then D_m = 0.148 × 10^{-7} cm²/s, that corresponds to experimental data for PVC [7]. The diffusivity coefficient D_n for nanocomposites calculation can be performed within the framework of multifractal model of gases diffusion according to the Equation (7):

$$D_n = D_m \left(\frac{\alpha_{ac}^n}{\alpha_{ac}^m} \right)^{d_m},$$

(10)

where, α_{ac}^n and α_{ac}^n are accessible for gases diffusion fractions of nanocomposite and matrix polymer, respectively, d_m is gas-diffusate molecule diameter, which is equal to 3.0 Å for O_2 [7].

The relative fraction of accessible for diffusion polymeric material α_{ac} is determined as follows [7]:

$$\alpha_{ac} = 1 - \phi_{cl} - \left(\phi_n + \phi_{if} \right), \tag{11}$$

where, φ_{cl}, φ_n, and φ_{if} are relative fractions of local order domains (nanoclusters), nanofiller, and interfacial regions, respectively, through which gases diffusion is not realized in virtue of their dense packing [7].

It is obvious, for matrix PVC plasticate $(\varphi_n + \varphi_{if}) = 0$. For nanocomposites PVC/MMT $(\varphi_n + \varphi_{if})$ value is determined with the aid of the following percolation relationship [6]:

$$\frac{E_n}{E_m} = 1 + 11 \left(\phi_n + \phi_{if} \right)^{1.7}, \tag{12}$$

where, E_n and E_m are elasticity moduli of nanocomposite and matrix polymer, respectively.

The value φ_{cl} is estimated as follows [8]:

$$\phi_{cl} = 0.03 \left(T_g - T \right)^{0.55}, \tag{13}$$

where, T_g and T are glass transition and tests temperatures, respectively. For PVC plasticate $T_g = 348$ K [4], that gives the value $\varphi_{cl} = 0.272$ for $T = 293$ K.

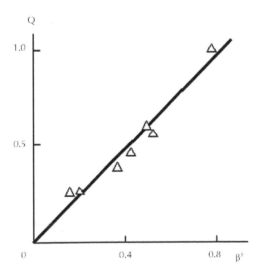

FIGURE 1 The dependence of combustion reaction completeness extent Q on exponent β for nanocomposites PVC/MMT.

Further the parameter β can be determined according to the Equation (2). In Figure 1 the relationship $Q(\beta^3)$ (such correlation type was chosen for its linearization) is adduced, which shows the strong dependence of material combustibility, expressed by the parameter Q, on the exponent β, characterizing combustion front instantaneous jumps ("Levy's flights") from one region of sample into another. So, β doubling results in fivefold nanocomposite volume, subjecting to combustion. The relationship $Q(\beta)$ is described analytically by the following empirical formula:

$$Q = 1.26\beta^3 . \tag{14}$$

Let us consider the structural basis of β change at organoclay contents variation. The distance between organoclay particles λ as the first approximation can be estimated according to the Equation (6):

$$\lambda = \left[\left(\frac{4\pi}{3\varphi_n} \right)^{1/3} - 2 \right] \left(\frac{D_p}{2} \right), \tag{15}$$

where, φ_n is organoclay volume content, D_p is nanofiller particle size, which for organoclay platelet can be estimated as proportional to cube root from its three main sizes product: length, thickness, and width, which are equal for exfoliated organoclay to 100, 1, and 35 nm, respectively [6].

The value φ_n can be determined according to the well-known Equation (6):

$$\varphi_n = \frac{W_n}{\rho_n}, \tag{16}$$

where, W_n is organoclay mass content, ρ_n is its density, which is estimated as follows (6):

$$\rho_n = 188(D_p)^{1/3}, \text{ kg/m3,} \tag{17}$$

where, D_p is given in nm.

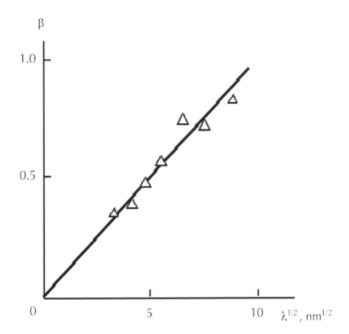

FIGURE 2   The dependence of exponent β on distance between nanofiller conditional particles λ for nanocomposites PVC/MMT.

In Figure 2 the dependence of β on $\lambda^{1/2}$ (such form of the indicated correlation was chosen again with its linearization purpose), which demonstrates β increasing at distance between organoclay particles enhancement and is described analytically by the following empirical equation:

$$\beta = 0.10\lambda^{1/2},\qquad(18)$$

where, λ is given in nm.

The Equation (14) and Equation (18) describe theoretically the experimental dependences of nanocomposites polymer/organoclay combustibility on nanofiller structure. So, organoclay contents φ_n increasing results in λ reduction and, respectively, in Q decrease. The transition from organoclay exfoliated structure up to intercalated one results in a packet (tactoid) from N organoclay platelets formation, that increases D_p value and, hence, enhances Q. Both indicated factors influence on β value: λ decreasing results in decay of combustion front jump ("Levy's flight") from one sample region into another, slowing thereby combustion process.

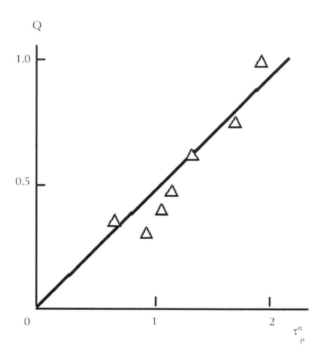

FIGURE 3 The dependence of combustion reaction completeness extent Q on fractal time τ_δ^α for nanocomposites PVC/MMT.

As it has been noted above the interval of active time, that is material combustion time, is proportional to t^α. This means, that reaction completeness degree Q can be determined as follows [7]:

$$Q \sim t^\alpha. \tag{19}$$

In Figure 3 the dependence of combustion reaction completeness degree Q on its fractal time τ_p^α for nanocomposites PVC/MMT is adduced. As it follows from this figure plot, the indicated dependence is approximated well enough by linear correlation that is it corresponds to the relationship (19) and is described by the following empirical equation:

$$Q = 0.47\tau_p^\alpha, \tag{20}$$

where τ_p is given in s.

The comparison with the similar formula for polymers synthesis [7] shows their principal distinction: the higher (in about 4 times) constant coefficient in the combustion case at the expence of process temperature enhancement. Let us note, that the exponents α and β are interconnected according to the Equation (2): combustion front jumps ("Levy's flights") intensity reinforcement results in α growth and, hence, to

fractal time τ_p^{α} enhancement and vica versa.

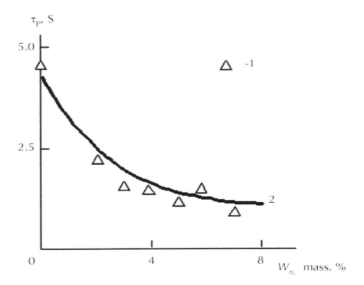

FIGURE 4 The experimental (1) and calculated according to the Equation (8) and Equation (10) and (2) dependences of putting out a fire time τ_p on organoclay mass contents W_n for nanocomposites PVC/MMT.

The Equation (8) and Equation (10) combination allows to estimate theoretically putting out a fire time τ_p. The theory and experiment comparison in the form of the dependence of τ_p on organoclay mass contents W_n (Figure 4) has shown their good correspondence.

CONCLUSION

Thus, the obtained in the present work results have shown usefulness of a strange (anomalous) diffusion conception for nanocomposites polymer/organoclay flame-resistance description. Organoclay platelets (tactoids) slow down combustion front jumps ("Levy's flights") from one sample region into another that is physical basis of the barrier effect. The combustion fractal time increasing results in "Levy's flights" intensification and vice versa. The proposed model describes correctly organoclay structure influence on nanocomposites flame-resistance.

KEYWORDS

- "Levy's flights"
- Flame-resistance
- Fractal Time.
- Nanocomposite
- Organoclay
- Strange Diffusion

REFERENCES

1. Lomakin, S. M. and Zaikov, G. E. *Vysokomolek. Soed. B* (in Russian.), **47**(1), 104–120 (2005).
2. Lomakin, S. M., Dubnikova, I. A., Berezina, S. M., and Zaikov, G. E. *Vysokomolek. Soed. A* (in Russion.), **48**(1), 90–105 (2006).
3. Clarey, M., Edwards, J., Tzipursky, S. J., Beall, G. W., and Eisenhour, D. D. Patent 6050509 USA (2001).
4. Khashirova, S. Yu., Borukaev, T. A., Sapaev, Kh. Kh., Ligidov, M. Kh., Kushkhov, Kh. B., and Mikitaev, A. K. Mater. VIII International Sci.-Pract. Conf. "New Polymer Composite Materials", *KBSU*, Nal'chik, 218–225 (2012).
5. Zelenyi, L. M. and Milovanov, A. V. *Uspekhi Fizicheskikh Nauk*, **174**(8), 809–852 (2004).
6. Mikitaev, A. K., Kozlov, G. V., and Zaikov, G. E. *Polymer Nanocomposites: Variety of Structural Forms and Applications*, Nova Science Publishers, Inc., New York, p. 319 (2008).
7. Kozlov, G. V., Zaikov, G. E., and Mikitaev, A. K. *The Fractal Analysis of Gas Transport in Polymers. The Theory and Practical Applications*, Nova Science Publishers, Inc., New York, p. 238 (2009).
8. Kozlov, G. V. and Zaikov, G. E. *Structure of the Polymer Amorphous State*. Brill Academic Publishers, Utrecht-Boston, p. 465 (2004).

CHAPTER 14

CALCULATION OF EFFICIENCY OF SEDIMENTATION OF DISPERSION PARTICLES IN A ROTOKLON ON THE BASIS OF MODEL OF HYDRODYNAMIC INTERACTING OF PHASES

R. R. USMANOVA and G. E. ZAIKOV

CONTENTS

14.1 INTRODUCTION

Common fault of the known wet-type collectors applied in industrial production, single-valued use of a fluid in dust removal process and, as consequence, its large charges on gas clearing is. For machining of great volumes of an irrigating liquid and slurry salvaging the facility of bulky, capital-intensive, difficult systems of water recycling which considerably process of clearing of gas and do a rise its commensurable with clearing cost at application of the most difficult and cost intensive systems of dry clearing of gases (electrostatic precepitators and bag hoses) is required.

In this connection necessity for creation of such wet-type collectors which would work with the low charge of an irrigating liquid now has matured and combined the basic virtues of modern means of clearing of gases: simplicity and compactness, a high performance, a capability of control of processes of a dust separation and optimisation of regimes.

To the greatest degree modern demands to the device and activity of apparatuses of clearing of industrial gases there match wet-type collectors with inner circulation the fluids gaining now more and more a wide circulation in systems of gas cleaning in Russia and abroad.

14.2 SURVEY OF KNOWN CONSTRUCTIONS OF SCRUBBERS WITH INNER CIRCULATION OF THE FLUID

An easy way to comply with the journal paper formatting requirements is to use this document as a template and simply type your text into it. The device and maintenance of systems of wet clearing of air are considerably facilitated, if water admission to contact zones implements as a result of its circulation in the apparatus. Slurry accumulating in it thus can continuously be retracted or periodically or by means of mechanical carriers, in this case necessity for water recycling system disappears, or a hydraulic path - a drain of a part of water. In the latter case the device of system of water recycling can appear expedient, but load on it is much less, than at circulation of all volume of water [1, 2].

Dust traps of such aspect are characterized by presence of the capacity filled with water. Cleared air contacts to this water, and contact conditions are determined by interacting of currents of air and waters. The same interacting calls a water circulation through a zone of a contact at the expense of energy of the most cleared air.

The water discharge is determined by its losses on transpiration and with deleted slurry. At slurry removal by mechanical scraper carriers or manually the water discharge minimum also makes only 2-5 g on 1 м³ air. At periodic drain of the condensed slurry the water discharge is determined by consistency of slurry and averages to 10 г on 1 м³ air, and at fixed drain the charge does not exceed 100-200 g on 1 м³ air. Filling of dust traps with water should be controlled automatically. Maintenance of a fixed level of water has primary value as its oscillations involve essential change as efficiency, and productivity of system.

The basic most known constructions of these apparatuses are introduced on Figure 1 [3].

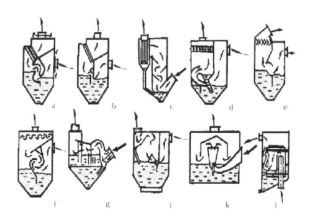

FIGURE 1   Constructions of scrubbers with inner circulation of a fluid: (a) - rotoklon N (USA); (b)- PVM CNII (Russia); (c) - a scrubber a VNIIMT (Russia);(d) - a dust trap to me (Czechoslovakia); (e) - dust trap WNA (Germany); (f) - dust trap "Asco» (Germany); (g) - dust trap LGP (Russia); (i) - dust trap "Klayrator» (USA); (k) - dust trap VDN (Austria); (l) - rotoklon RPA a NIIOGAS (Russia).

Mechanically each of such apparatuses consists of contact channel fractionally entrained in a fluid and the drip pan merged in one body. The principle of act of apparatuses is grounded on a way of intensive wash down of gases in contact channels of a various configuration with the subsequent separation of a water gas flow in the drip pan. The fluid which has thus reacted and separated from gas is not deleted at once from the apparatus, and circulates in it and is multiply used in dust removal process.

Circulation of a fluid in the wet-type collector is supplied at the expense of a kinetic energy of a gas flow. Each apparatus is supplied by the device for maintenance of a fixed level of a fluid, and also the device for removal of slurry from the scrubber collecting hopper.

Distinctive features of apparatuses:

1. Irrigating of gas by a fluid without use of injectors that allows to use for irrigating a fluid with the high contents of suspended matters (to 250 mg/m³);
2. Landlocked circulation of a fluid in apparatuses which allows to reuse a fluid in contact devices of scrubbers and by that to device out its charge on clearing of gas to 0,5kg/m³, i.e. in 10 and more times in comparison with other types of wet-type collectors;
3. Removal of a collected dust from apparatuses in the form of dense шламов with low humidity that allows to simplify dust salvaging to diminish load by water treating systems, and in certain cases in general to refuse their facility;
4. Layout of the drip pan in a body of the apparatus which allows to diminish sizes of dust traps to supply their compactness.

The indicated features and advantages of such scrubbers have led to wide popularity of these apparatuses, active working out of various constructions, research and a heading of wet-type collectors, as in Russia, and abroad.

The scrubbers introduced on Figure 1. Concern to apparatuses with non-controllable operating conditions as in them there are no gears of regulating. In scrubbers of this type the stable conditions of activity of a high performance is difficultly supplied, especially at varying parameters of cleared gas (pressure, temperature, a volume, a dust content and т.д). In this connection wet scrubbers with controlled variables are safer and perspective. Regulating of operating conditions allows to change a hydraulic resistance from which magnitude, according to the power theory of a wet dust separation, efficiency of trapping of a dust depends. Regulating of parameters allows to operate dust traps in an optimum regime at which optimum conditions of interacting of phases are supplied and peak efficiency of trapping of a dust with the least power expenditures is attained. The great value is acquired by dust traps with adjustable resistance also for stabilization of processes of gas cleaning at varying parameters of cleared gas. A row of such scrubbers is introduced in Figure 2.

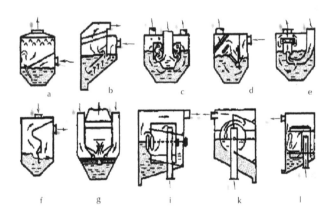

FIGURE 2   Apparatuses with controlled variables: (a) - under the patent №1546651 (Germany), (b) - the ACE №556824 (USSR), (c) - the ACE № 598625 (USSR), (d) - the ACE №573175 (USSR), (e) - under the patent № 1903985 (Germany), (f) - the ACE № 13686450 (France), (g)- the ACE № 332845 (USSR), (i) - *the* ACE № 318402 (USSR), (k) - the ACE № 385598 (USSR), (l) - type RPA a NIIOGAS (USSR).

In the scrubbers introduced on Figure 2, and, turn of controlling partitions is made either manually, or is distant with the drive from the electric motor, and in a dust trap on Figure 2, in - manually, moving of partitions rather each other on a threaded connection. In dust traps on Figure 2 (e). The lower partitions are executed in the form of a float or linked with floats that allow to stabilize clearing process at varying level of a fluid.

The interesting principle of regulating is applied in the dust traps figured on Figure 1 and Figure 2. In these apparatuses contact devices are had on a wall of the floating chamber entrained in a fluid and hardened in a body by means of joints. Such construction of dust traps allows to support automatically to constants an apparatus hydraulic resistance at varying gas load.

From literary data follows that known constructions of scrubbers with inner circulation of a fluid work in a narrow range of change of speed of gas in contact channels and are used in industrial production in the core for clearing of gases of a size dispersivity dust in systems of an aspiration of auxiliaries [3-5]. Known apparatuses are rather sensitive to change of gas load on the contact channel and to fluid level, negligible aberrations of these parameters from best values lead to a swing of levels of a fluid at contact channels, to unstablis operational mode and dust clearing efficiency lowering. Because of low speeds of gas in contact channels known apparatuses have large gabarits. These deficiencies, and also a weak level of scrutiny of processes proceeding in apparatuses, absence of safe methods of their calculation hamper working out of new rational constructions of wet-type collectors of the given type and their wide heading in manufacture. In this connection necessity of more detailed theoretical and experimental study of scrubbers with inner circulation of a fluid for the purpose of the prompt use of the most effective and costeffective constructions in systems of clearing of industrial gases has matured.

14.3   ARCHITECTURE OF HYDRODYNAMIC INTERACTING OF PHASES

In scrubbers with inner circulation of a fluid process of interacting of gas, liquid and hard phases in which result the hard phase (dust), finely divided in gas, passes in a fluid implements. Because density of a hard phase in gas has rather low magnitudes (to 50 g/m^3), it does not render essential agency on hydrodynamics of flows. Thus, hydrodynamics study in a scrubber with inner circulation of a fluid is reduced to consideration of interacting of gas and liquid phases.

The process of hydrodynamic interacting of phases it is possible to disjoint sequentially proceeding stages on the following:
- Fluid acquisition by a gas flow on an entry in the contact device:
- Fluid subdivision by a fast-track gas flow in the contact channel;
- Integration of drops of a fluid on an exit from the contact device;
- Branch of drops of a fluid from gas in the drip pan.

14.3.1   FLUID ACQUISITION BY A GAS FLOW ON AN ENTRY IN THE CONTACT DEVICE

Before an entry in the contact device of the apparatus there is a cotraction of a gas flow to increase in its speed, acquisition of high layers of a fluid and its hobby in the contact channel. Functionability of all dust trap depends on efficiency of acquisition of a fluid a gas flow - without fluid acquisition will not be supplied effective interacting of phases in the contact channel and, hence, qualitative clearing of gas of a dust will not be attained. Thus, fluid acquisition by a gas flow on an entry in the contact device is one of defined stages of hydrodynamic process in a scrubber with inner circulation of a fluid. Fluid acquisition by a gas flow can be explained presence of interphase turbulence which is advanced on an interface of gas and liquid phases. A condition for origination of interphase turbulence is presence of a gradient of speeds of phases on boundaries, difference of viscosity of flows, an interphase surface tension. At gas driving over a surface of a fluid the last will break gas boundary layers therefore in them there are the turbulent shearing stresses promoting cross-section transfer of energy. Originating cross-section turbulent oscillations lead to penetration of turbulent gas curls into boundary layers of a fluid with the subsequent illuviation of these stratums in curls. Mutual penetration of curls of boundary layers leads as though to the clutch of gas with a fluid on a phase boundary and to hobby of high layers of a fluid for moving gas over its surface. Intensity of such hobby depends on a kinetic energy of a gas flow, from its speed over a fluid at an entry in the contact device. At gradual increase in speed of gas there is a change of a surface of a fluid at first from smooth to undular, then ripples are organized and, at last, there is a fluid dispersion in gas. Mutual penetration of curls of boundary layers leads as though to the clutch of gas with a fluid on a phase boundary and to hobby of high layers of a fluid for moving gas over its surface. Intensity of such hobby depends on a kinetic energy of a gas flow, from its speed over a fluid at an entry in the contact device. The quantitative assessment of efficiency of acquisition in wet-type collectors with inner circulation of a fluid is expedient for conducting by means of a parameter $m = V_z/V_g$ m³/m³ equal to a ratio of volumes of liquid and gas phases in contact channels and characterizing the specific charge of a fluid on gas irrigating in channels. Obviously that magnitude m will be determined, first of all, by speed of a gas flow on an entry in the contact channel. Other diagnostic variable is fluid level on an entry in the contact channel which can change cross-section of the channel and influence speed of gas:

$$\frac{v_r}{s_r} = \frac{V_r}{bh_k - bh_g} - \frac{v_r}{b(h_k - h_g)} \tag{1}$$

where, S_r - cross-section of the contact channel; b - a channel width; h_K - channel altitude; h_g - fluid level.

Thus, for the exposition of acquisition of a fluid a gas flow in contact channels it is enough to gain experimental relation of following type:

$$m = f(W_r, h_æ) \tag{2}$$

14.3.2 FLUID SUBDIVISION BY A FAST-TRACK GAS FLOW IN THE CONTACT CHANNEL

As shown further, efficiency of trapping of corpuscles of a dust in many respects depends on a size of drops of a fluid: with decrease of a size of drops the dust clearing efficiency raises. Thus, the given stage of hydrodynamic interacting of phases is rather important.

The process of subdivision of a fluid by a gas flow in the contact channel of a dust trap occurs at the expense of high relative speeds between a fluid and a gas flow. For calculation of average diameter of the drops gained in contact channels, it is expedient to use the empirical formula of the Japanese engineers Nukiymas and Tanasavas which allows to consider agency of operating conditions along with physical performances of phases:

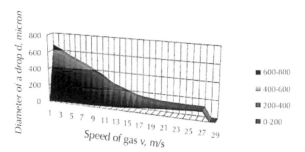

FIGURE 3   Relation of an average size of drops of water in blade impellers from speed of gas.

$$D_o = \frac{585 \cdot 10^3 \sqrt{\sigma}}{W_r} + 49{,}7 \left(\frac{\mu_l}{\sqrt{\rho_l \sigma_l}} \right)^{0,2} \frac{L_l}{V_r} \tag{3}$$

where, W_r - relative speed of gases in the channel, m/s; σ_l - factor of a surface tension of a fluid, N/m; ρ_l - fluid density, kg/m³; μ_l - viscosity of a fluid, the Pas/with; L_l - volume-flow of a fluid, m3/with; V_r - volume-flow of gas, m³/with.

In Figure 3 computational curves of average diameter of drops of water in contact channels depending on speed of a gas flow are resulted. Calculation is conducted by formula (3) at following values of parametres: $\sigma = 720 \cdot 10^3$ N/m; $\rho_l = 1000$ kg/m³; $\mu = 1{,}01 \cdot 10^2$ P/s.

The gained relations testify that the major operating conditions on which the average size of drops in contact channels depends, speed of gas flow W_r and the specific charge of a fluid on gas irrigating m are. These parametres determine hydrodynamic structure of an organized water gas flow.

With growth of speed of gas process of subdivision of a fluid by a gas flow gains in strength, and drops of smaller diameter are organized. The most intensive agency on a size of drops renders change of speed of gas in the range from 7 to 20 m/s, at the further increase in speed of gas (> 20 m/s) intensity of subdivision of drops is reduced. It is necessary to note that in the most widespread constructions of shock-inertial apparatuses (rotoklons N) which work at speed of gas in contact devices of 15 m/s, the size of drops in the channel is significant and makes 325-425 microns. At these operating conditions and sizes of drops qualitative clearing of gas of a mesh dispersivity dust is not attained. For decrease of a size of drops and raise of an overall performance of these apparatuses the increase in speed of gas to 30, 40, 50 m/s and more depending on type of a trapped dust is necessary.

The increase in the specific charge of a fluid at gas irrigating leads to growth of diameter of organized drops. So, at increase m with $0{,}1 \cdot 10^3$ to $3 \cdot 10$ m³/m³ the average size of drops is increased approximately at 150 microns. For security of minimum diameter of drops in contact channels of shock-inertial apparatuses the specific charge of a fluid on gas irrigating should be optimized over the range $(0,1-1,5 \cdot 10\ )m^3/m^3$. It is necessary to note that in the given range of specific charges with a high performance the majority of fast-track wet-type collectors work.

14.3.3 INTEGRATION OF DROPS OF A FLUID ON AN EXIT FROM THE CONTACT DEVICE

On an exit from the contact device there is an expansion of a water gas stream and integration of drops of a fluid at the expense of their concretion. The maximum size of the drops weighed in a gas flow, is determined by stability conditions: the size of drops will be that more than less speed of a gas flow. Thus, on an exit from the contact device together with fall of speed of a gas flow the increase in a size of drops will be observed. Turbulence in an extending part of a flow more than in the channel with fixed cross-section, and it grows with increase in an angle of jet divergence, and it means that speed of turbulent concretion will grow in an extending part of a flow also with increase in an angle of jet divergence. The more full there will be a concretion of corpuscles of a fluid, the drop on an exit from the contact device will be larger and the more effectively they will be trapped in the drip pan.

Practice shows that the size a coagulation of drops on an exit makes of the contact device, as a rule, more than 150 microns. Corpuscles of such size are easily trapped in the elementary devices (the inertia, gravitational, centrifugal, etc.).

14.3.4 BRANCH OF DROPS OF A FLUID FROM A GAS FLOW

The inertia and centrifugal drip pans are applied to branch of drops of a fluid from gas in shock-inertial apparatuses in the core. In the inertia drip pans the branch implements at the expense of veering of a water gas flow. Liquid drops, moving in a gas flow, possess definitely a kinetic energy thanks to which at veering of a gas stream they by inertia move rectilinearly and are inferred from a flow. If to accept that the drop is in the form of a sphere and speed of its driving is equal in a gas flow to speed of this flow the kinetic energy of a drop, moving in a flow, can be determined by formula:

$$E_k = \frac{\pi D_0^{\ 3}}{6} \rho_l \frac{W_r^2}{2} \tag{4}$$

with decrease of diameter of a drop and speed of a gas flow the drop kinetic energy is sharply diminished. At gas-flow deflection the inertial force forces to move a drop in a former direction. The more the drop kinetic energy, the is more and an inertial force:

$$E_k = \frac{\pi D_0^{\ 3}}{6} \rho_l \frac{dW_r}{d\tau} \tag{5}$$

Thus, with flow velocity decrease in the inertia drip pan and diameter of a drop the drop kinetic energy is diminished, and efficiency drop spreads is reduced. However the increase in speed of a gas flow cannot be boundless as in a certain velocity band of gases there is a sharp lowering of efficiency drop spreads owing to origination of secondary ablation the fluids trapped drops. For calculation of a breakdown speed of gases in the inertia drip pans it is possible to use the formula, m/s:

$$W_c = K \sqrt{\frac{\rho_l - \rho_e}{\rho_r}} \qquad (6)$$

where, W_c - optimum speed of gases in free cross-section of the drip pan, m/s; K - the factor defined experimentally for each aspect of the drip pan.

Values of factor normally fluctuate over the range 0,1–0,3. Optimum speed makes from 3 to 5 m/s.

14.4   PURPOSE AND RESEARCH PROBLEMS

The following was the primal problems which were put by working out of a new construction of the wet-type collector with inner circulation of a fluid:

- creation of a dust trap with a broad band of change of operating conditions and a wide area of application, including for clearing of gases of the basic industrial assemblies of a mesh dispersivity dust;
- creation of the apparatus with the operated hydrodynamics, allowing to optimise process of clearing of gases taking into account performances of trapped ingredients;
- to make the analysis of hydraulic losses in blade impellers and to state a comparative estimation of various constructions of contact channels of an impeller by efficiency of security by them of hydrodynamic interacting of phases;
- to determine relation of efficiency of trapping of corpuscles of a dust in a rotoklon from performance of a trapped dust and operating conditions major of which is speed of a gas flow in blade impellers. To develop a method of calculation of a dust clearing efficiency in scrubbers with inner circulation of a fluid;
- definition of boundary densities of suspension various a dust after which excess general efficiency of a dust separation is reduced;
- definition of the maximum extent of circulation of an irrigating liquid;
- to gain the differential equation of driving of corpuscles with which help was possiblly to determine paths of their driving in the field of a leak-in of a gas flow on a fluid surface, and also to count limiting sizes of corpuscles which can be precipitated on a fluid surface.

14.5   EXPERIMENTAL RESEARCHES

14.5.1   THE EXPOSITION OF EXPERIMENTAL INSTALLATION AND THE TECHNIQUE OF REALIZATION OF EXPERIMENT

The rotoklon represents the basin with water on which surface on a connecting pipe of feeding into of dusty gas the dust-laden gas mix arrives. Over a water surface gas deploys, and a dust contained in gas by inertia penetrate into a fluid. Turn of blades of an impeller is made manually, rather each other on a threaded connection by means of handwheels. The slope of blades was installed in the interval 25°–45° to an axis.

In a rotoklon three pairs lobes sinusoidal a profile, the regulatings of their rule executed with a capability are installed. Depending on cleanliness level of an airborne dust flow the lower lobes by means of handwheels are installed on an angle defined by operational mode of the device. The rotoklon is characterised by presence of three slotted channels, a formation the overhead and lower lobes, and in everyone the subsequent on a course of gas the channel the lower lobe is installed above the previous. Such arrangement promotes a gradual entry of a water gas flow in slotted channels and reduces thereby a device hydraulic resistance. The arrangement of an input part of lobes on an axis with a capability of their turn allows to create a diffusion reacting region. Sequentially had slotted channels create in a diffusion zone organised by a turn angle of lobes, a hydrodynamic zone of intensive wetting of corpuscles of a dust. In process of flow moving through the fluid-flow curtain, the capability of multiple stay of corpuscles of a dust in hydrodinamically reacting region is supplied that considerably raises a dust clearing efficiency and ensures functioning of the device in broad bands of cleanliness level of a gas flow.

The construction of a rotoklon with adjustable sinusoidal lobes is developed and protected by the patent of the Russian Federation, capable to solve a problem of effective separation of a dust from a gas flow [6]. Thus water admission to contact zones implements as a result of its circulation in the apparatus.

The rotoklon with the adjustable sinusoidal lobes, introduced on Figure 4 contains a body 3 with connecting pipes for an entry 7 and an exit 5 gases in which steams of lobes sinusoidal a profile are installed. Moving of the overhead lobes 2 implements by means of screw jacks 6, the lower lobes 1 are fixed on an axis 8 with a capability of their turn. The turn angle of the lower lobes is chosen from a condition of a persistence of speeds of an airborne dust flow. For regulating of a turn angle output parts of the lower lobes 1 are envisioned handwheels. Quantity of pairs lobes is determined by productivity of the device and cleanliness level of an airborne dust flow, that is a regime of a stable running of the device. In the lower part of a body there is a connecting pipe for a drain of slime water 9. Before a connecting pipe for a gas make 5 the labyrinth drip pan 4 is installed. The rotoklon works as follows. Depending on cleanliness level of an airborne dust flow the overhead lobes 5 by means of screw jacks 6, and the lower lobes 1 by means of handwheels are installed on an angle defined by operational mode of the device. Dusty gas arrives in the upstream end 7 in a top of a body 3 apparatuses. Hitting about a fluid surface, it changes the direction and passes in the slotted channel organised overhead 2 and lower 1 lobes. Thanks to the driving high speed, cleared gas captures a high layer of a fluid and atomises it in the smallest drops and foam with an advanced surface. After consecutive transiting of all slotted channels gas passes through the labyrinth drip pan 4 and through the discharge connection 5 is deleted in an aerosphere. The collected dust settles out in the loading pocket of a rotoklon and through a connecting pipe for a drain of slime water 9, together with a fluid, is periodically inferred from the apparatus.

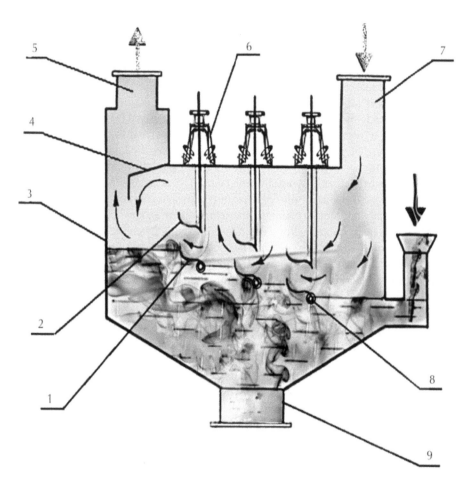

FIGURE 4 A rotoklon General view: Lower 1 and the overhead 2 lobes; a body 3; the labyrinth drip pan 4; connecting pipes for an entry 7 and an exit 5 gases; screw jacks 6; an axis 8; a connecting pipe for a drain of slurry 9.

Noted structural features does not allow to use correctly available solutions on hydrodynamics of dust-laden gas flows for a designed construction. In this connection, for the well-founded exposition of the processes occurring in the apparatus, there was a necessity of realisation of experimental researches.

Experiments were conducted on the laboratory-scale plant "rotoklon" introduced on Figure 5.

The examined rotoklon had 3 slotted channels speed of gas in which made to 15 km/s. At this speed the rotoklon had a hydraulic resistance 800 Pases. Working in such regime, it supplied efficiency of trapping of a dust with input density 0,5 g/nm³ and density 1200 kg/m³ at level of 96,3% [7].

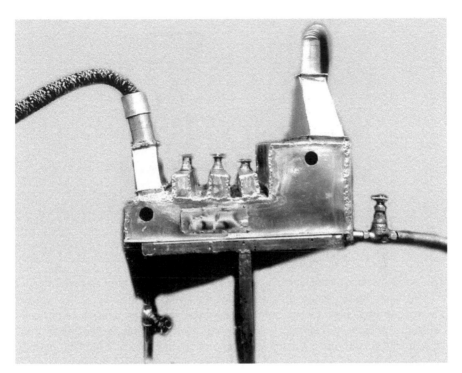

FIGURE 5   Experimental installation "rotoklon".

In the capacity of modelling system air and a dust of talc with a size of corpuscles d = 2 ÷30 a micron, white black and a chalk have been used. The apparatus body was filled with water on level h_g = 0,175m.

Cleanliness level of an airborne dust mix was determined by a direct method [8]. On direct sections of the pipeline before and after the apparatus the mechanical sampling of an airborne dust mix was made. After determination of matching operational mode of the apparatus, gas test were taken by means of intaking handsets. Mechanical sampling isocinetycs on intaking handsets were applied to observance replaceable tips of various diameters. Full trapping of the dust contained in taken test of an airborne dust mix, was made by an external filtering draws through mixes with the help calibrates electroaspirator EA-55 through special analytical filters AFA-10 which were put in into filtrating cartridges. The selection time was fixed on a stop watch, and speed - the rotameter of electroaspirator EA-55.

Dust gas mix gained by dust injection in the flue by means of the metering screw conveyer batcher introduced on Figure 6. Application of the batcher with varying productivity has given the chance to gain the set dust load on an entry in the apparatus.

FIGURE 6   The metering screw conveyer batcher of a dust.

The water discharge is determined by its losses on transpiration and with deleted slurry. The water drain is made in the small portions from the loading pocket supplied with a pressure lock. Gate closing implements sweeping recompression of air in the gate chamber, opening–a depressurization. Small level recession is sweepingly compensated by a top up through a connecting pipe of feeding into of a fluid. At periodic drain of the condensed slurry the water discharge is determined by consistency of slurry and averages to 10 г on 1 м³ air, and at fixed drain the charge does not exceed 100–200 г on 1 м³ air. Filling of a rotoklon with water was controlled by means of the level detector. Maintenance of a fixed level of water has essential value as its oscillations involve appreciable change as efficiency, and productivity of the device.

14.6   DISCUSSION AND RESULTS

In a rotoklon process of interacting of gas, liquid and hard phases in which result the hard phase (dust), finely divided in gas, passes in a fluid is realized. Process of hydrodynamic interacting of phases in the apparatus it is possible to disjoint sequentially proceeding stages on the following: fluid acquisition by a gas flow on an entry in the contact device; fluid subdivision by a fast-track gas flow in the contact channel; concretion of dispersion particles by liquid drops; branch of drops of a fluid from gas in the labyrinth drip pan.

At observation through an observation port the impression is made that all channel is filled by foam and water splashes. Actually this effect caused by a retardation of a flow at an end wall, is characteristic only for a stratum which directly is bordering on to glass. Slow-motion shot consideration allows to install a true flow pattern. It is visible that the air jet as though itself chooses the path, being aimed to be punched in the shortest way through water. Blades standing sequentially under existing conditions restrict air jet extending, forcing it to make sharper turn that, undoubtedly, favors to separation. Functionability of all dust trap depends on efficiency of acquisition of a fluid a gas flow - without fluid acquisition will not be supplied effective interacting of phases in contact channels and, hence, qualitative clearing of gas of a dust will not be attained. Thus, fluid acquisition by a gas flow at consecutive transiting of blades of an impeller is one of defined stages of hydrodynamic process in a rotoklon.

Fluid acquisition by a gas flow can be explained presence of interphase turbulence which is advanced on an interface of gas and liquid phases. A condition for origination of interphase turbulence is presence of a gradient of speeds of phases on boundaries, difference of viscosity of flows, an interphase surface tension.

14.6.1 THE ESTIMATION OF EFFICIENCY OF GAS CLEANING

The quantitative assessment of efficiency of acquisition in apparatuses of shock-inertial type with inner circulation of a fluid is expedient for conducting by means of a parameter $n = L_l/L_g$, m^3/m^3 equal to a ratio of volumes of liquid and gas phases in contact channels and characterizing the specific charge of a fluid on gas irrigating in channels. Obviously that magnitude n will be determined, first of all, by speed of a gas flow on an entry in the contact channel. The following important parameter is fluid level on an entry in the contact channel which can change cross-section of the channel and influence speed of gas:

$$\frac{\vartheta_g}{S_g} = \frac{\vartheta_g}{bh_k - bh_l} - \frac{\vartheta_g}{b(h_k - h_l)} \tag{7}$$

where, S_g - cross-section of the contact channel; b - a channel width; h_K - channel altitude; h_l - fluid level.

Thus, for the exposition of acquisition of a fluid a gas flow in contact channels of a rotoklon it is enough to gain the following relation experimentally:

$$n = f(\vartheta_g \cdot h_l) \tag{8}$$

As it has been installed experimentally, efficiency of trapping of corpuscles of a dust in many respects depends on a size of drops of a fluid: with decrease of a size of drops the dust clearing efficiency raises. Thus, the given stage of hydrodynamic interacting of phases is rather important. For calculation of average diameter of the drops organised at transiting of blades of an impeller, the empirical relation is gained:

$$d = \frac{467 \cdot 10^3 \sqrt{\sigma}}{\vartheta_o} + 17,869 \cdot \left(\frac{\mu_l}{\sqrt{\rho_l \sigma}}\right)^{0,68} \frac{L_l}{L_r} \tag{9}$$

where, ϑ_l - relative speed of gases in the channel, m/s; σ - factor of a surface tension of a fluid, N/m; ρ_1 - fluid density, kg/m^3; μ_l - viscosity of a fluid, the Pas/with; L_l - volume-flow of a fluid, m^3/with; L_g - volume-flow of gas, m^3/with.

The offered formula allows to consider also together with physical performances of phases and agency of operating conditions.

In Figure 7 Design values of average diameter of the drops organised at transiting of blades of an impeller, from speed of gas in contact channels and a gas specific irrigation are introduced. At calculation values of physical properties of water were accepted at temperature 20°C: $\rho_1 = 998$ kg/m³; $\mu_1 = 1,002 \cdot 10^{-3}$ N · C/m², $\varsigma = 72,86$ $\cdot 10^{-3}$ N/m.

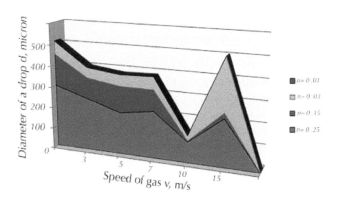

FIGURE 7   Computational relation of a size of drops to flow velocity and a specific irrigation.

The gained relations testify that the major operating conditions on which the average size of drops in contact channels of a rotoklon depends, speed of gas flow ϑ_i and the specific charge of a fluid on gas irrigating n are. These parameters determine hydrodynamic structure of an organized water gas flow.

Separation efficiency of gas bursts in apparatuses of shock-inertial act can be discovered only on the basis of empirical data on particular constructions of apparatuses. Methods of the calculations, found application in projection practice, are grounded on an assumption about a capability of linear approximation of relation of separation efficiency from diameter of corpuscles in is likelihood-logarithmic axes. Calculations on a likelihood method are executed under the same circuit design, as for apparatuses of dry clearing of gases [9].

Shock-inertial sedimentation of corpuscles of a dust occurs at flow of drops of a fluid by a dusty flow therefore the corpuscles possessing inertia, continue to move across the curved stream-lines of gases, the surface of drops attain and are precipitated on them.

Efficiency of shock-inertial sedimentation η_u is function of following dimensionless criterion:

$$\eta_{\grave{e}} = f\left(\frac{m_p}{\xi_c} \cdot \frac{\vartheta_p}{d_0}\right) \tag{10}$$

where, m_p - mass of a precipitated corpuscle; ϑ_p - speed of a corpuscle; ξ - factor of resistance of driving of a corpuscle; d_0 - diameter a midelev of cross-section of a drop.

For the spherical corpuscles which driving obeys the law the Stokes, this criterion looks like the following:

$$\frac{m_p \vartheta_p}{\xi_c d_0} = \frac{1}{18} \cdot \frac{d_r^2 \vartheta_p \rho_p C_c}{\mu_g d_0} \tag{11}$$

Complex $d_p^2 \vartheta_p \rho_p C_c / (18 \mu_g d_0)$ is parametre (number) of the Stokes:

$$\eta_{\dot{e}} = f(Stk) = f\left(\frac{d_p^2 \vartheta_p \rho_p C_c}{18 \mu_g d_0}\right) \tag{12}$$

Thus, efficiency of trapping of corpuscles of a dust in a rotoklon on the inertia model depends primarily on performance of a trapped dust (a size and density of trapped corpuscles) and operating conditions major of which is speed of a gas flow at transiting through blades of impellers.

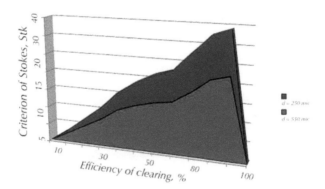

FIGURE 8   Relation of efficiency of clearing of gas to criterion StK.

On the basis of the observed inertia of model the method of calculation of a dust clearing efficiency in scrubbers with inner circulation of a fluid is developed.

The basis for calculation on this model is the formula (12). For calculation realization it is necessary to know disperse composition of a dust, density of corpuscles of a dust, viscosity of gas, speed of gas in the contact channel and the specific charge of a fluid on gas irrigating.

Calculation is conducted in the following sequence:

- by formula (9) determine an average size of drops D_0 in the contact channel at various operating conditions;
- by formula (10) count the inertia parametre of the Stokes for each fraction of a dust;
- by formula (11) fractional values of efficiency η for each fraction of a dust;
- general efficiency of a dust separation determine by formula (12), %.

The observed inertia model full enough characterises physics of the process proceeding in contact channels of a rotoklon.

14.6.2   COMPARISON OF EXPERIMENTAL AND COMPUTATIONAL RESULTS

Analyzing the gained results of researches of general efficiency of a dust separation, it is necessary to underscore that in a starting phase of activity of a dust trap for all used in researches a dust separation high performances, components from 93,2% for carbon black to 99,8 % for a talc dust are gained. Difference of general efficiency of trapping of various types of a dust originates because of their various particle size distributions on an entry in the apparatus, and also because of the various form of corpuscles, their dynamic wettability and density. The gained high values of general efficiency of a dust separation testify to correct selection of constructional and operation parameters of the studied apparatus and indicate its suitability for use in engineering of a wet dust separation.

As appears from introduced in Figure 9–10 graphs, the relation of general efficiency of a dust separation to speed of a mixed gas and fluid level in the apparatus will well be agreed to design data that confirms an acceptability of the accepted assumptions.

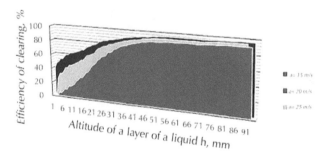

FIGURE 9   Relation of efficiency of clearing of gas to irrigating liquid level.

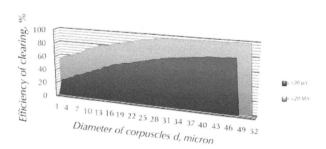

FIGURE 10   Dependence of efficiency of clearing of gas on a size of corpuscles and speed of gas.

In Figure 11, Figure 12, and Figure 13 results of researches on trapping various a dust in a rotoklon with adjustable sinusoidal are shown. The given researches testify to a high performance of trapping of corpuscles of thin a dust with their various moistening ability. From these drawings by fractional efficiency of trapping it is obviously visible, what even for corpuscles a size less than 1 microns (which are most difficultly trapped in any types of dust traps) к.п.д. Installations considerably above 90%. Even for the unwettable sewed type of white black general efficiency of trapping more than 96%. Naturally, as for the given dust trap lowering of fractional efficiency of trapping at decrease of sizes of corpuscles less than 5 microns, however not such sharp, as for other types of dust traps is characteristic.

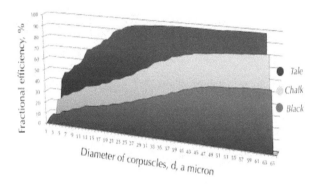

FIGURE 11   Fractional efficiency of clearing of corpuscles of a various dust.

CONCLUSIONS

1. The new construction of the rotoklon is developed, allowing to solve a problem of effective separation of a dust from a gas flow. In the introduced apparatus water admission to contact zones implements as a result of its circulation in the device.

2. Experimentally it is shown that fluid acquisition by a gas flow at consecutive transiting of blades of an impeller is one of defined stages of hydrodynamic process in a rotoklon.

3. Are theoretically gained and confirmed by data of immediate measurements of value of efficiency of shock-inertial sedimentation of dispersion particles in a rotoklon. The gained computational relationships, allow to size up the contribution as performances of a trapped dust (a size and density of trapped corpuscles), and operating conditions major of which is speed of a gas flow at transiting through blades of impellers.

4. Good convergence of results of scalings on the gained relationships with the data which are available in the technical literature and own experiments confirms an acceptability of the accepted assumptions.

The formulated leading-outs are actual for intensive operation wet-type collectors in which the basic gear of selection of corpuscles is the gear of the inertia dust separation.

KEYWORDS

- **A particle path**
- **An irrigating liquid**
- **Blade impellers**
- **Dust clearing efficiency**
- **Rotoclon**
- **Shock-inertial**

REFERENCES

1. Uzhov, V. N., Valdberg, A. J., and Myagkov, B. I. *Clearing of industrial gases of a dust*, Moscow: Chemistry (1981).
2. Pirumov, A. I. *air Dust removal*, Engineering industry, Moscow (1974).
3. Shvydky, V. S. and Ladygichev, M. G. *Clearing of gases. The directory*, Heat power engineering, Moscow (2002).
4. Straus, V. *Industrial clearing of gases*, Chemistry, Moscow (1981).
5. Kouzov, P. A., Malgin, A. D., and Skryabin, G. M. *Clearing of gases and air of a dust in the chemical industry*. Chemistry, St.-Petersburg (1993).
6. Patent 2317845 RF, IPC, cl. B01 D47/06 *Rotoklon a controlled sinusoidal blades*. Usmanova, R. R., Zhernakov, V. S., and Panov, A. K. Publ. 27.02.2008. Bull. № 6.
7. Usmanova, R. R., Zaikov, G. E., Stoyanov, O V., and Klodziuska, E. *Research of the mechanism of shock-inertial deposition of dispersed particles from gas flow the* bulletin of the Kazan technological university №9, pp. 203–207 (2013).
8. Kouzov, P. A. *Bases of the analysis of disperse composition industrial a dust*. Chemistry, Leningrad, pp. 183–195 (1987).
9. Vatin, N. I. and Strelets, K. I. *Air purification by means of apparatuses of type the cyclone separator*. St.-Petersburg (2003).

CHAPTER 15

HYALURONAN–A HARBINGER OF THE STATUS AND FUNCTIONALITY OF THE JOINT

LADISLAV ŠOLTÉS and GRIGORIJ KOGAN

CONTENTS

15.1 INTRODUCTION

Women often live longer than men. This fact could be associated with their enhanced redox load during the reproductive phase of their life. The physiological bleeding (with a periodicity of ca. 4 weeks) is accompanied by changes in the concentration of iron ions. Pre-menopausal women are believed to have a lower risk of common diseases because amounts of iron in their body are unlikely to be excessive at this time [1].

Fe ions are regarded as one of the most important catalytical agents that contribute to the augmented generation of the reactive oxygen species (for example, •OH radicals). However, such "radical training" of female organism lasting on average 40 yr (that is in a period between ca. 15 to 55 yr) can have a positive effect on females in the sense that their organism is better adjusted to the oxidative stress. In the "free radical theory of ageing" oxidative stress is considered to be a risk factor that is usually associated with such negative consequences as serious diseases or even premature death [2,3].

Life can be in a simplified way divided into three periods: childhood, maturity, and senescence. Maturity is the longest lasting part of human life. It lasts from the end of development and growth of a skeleton (around ca. 20 yr) till the old age, which start can be marked as at ca. 70–75 yr. Thus, maturity lasts about half a century. During this period, human skeleton can be considered invariable regarding the number of bones (206), their size, and mass.

The human skeleton consists of both fused and individual bones supported and supplemented by ligaments, tendons, and skeletal muscles. The articular ligaments and tendons are the main parts holding together the joint(s). In respect to the movement, there are freely moveable, partially moveable, and immovable joints. Synovial joints, the freely moveable ones, allow for a large range of motion and encompass wrists, knees, ankles, shoulders, and hips.

15.2 THE STRUCTURE OF A SYNOVIAL JOINT

15.2.1 CARTILAGE

In a healthy synovial joint, heads of the bones are encased in a smooth (hyaline) cartilage layer. These tough slippery layers – for example, those covering the bone ends in the knee joint – belong to mechanically highly stressed tissues in the human body. At walking, running, or sprinting the strokes frequency attain approximately 0.5, 2.5 or up to 10 Hz.

Figure 1 illustrates a normal healthy synovial joint indicating its major parts

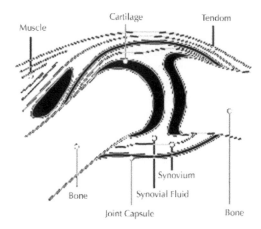

FIGURE 1 Normal, healthy synovial joint [4].z.e The cartilage functions also as a shock absorber. This property is derived from. its high water-entrapping capacity, as well as from the structure and intermolecular interactions among polymeric components that constitute the cartilage tissue [5]. Figure 2 sketches a section of the cartilae– a chondrocyte cell that permanently restructures/rebuilds its extracellular matri.

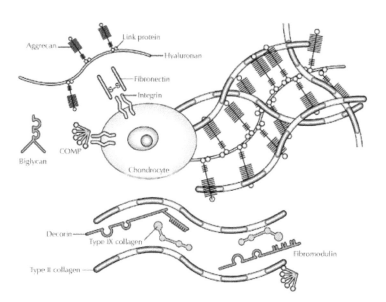

FIGURE 2   Articular cartilage main components and structure [6].

Three classes of proteins exist in articular cartilage: collagens (mostly type II collagen); proteoglycans (primarily aggrecan); and other noncollagenous proteins (including link protein, fibronectin, COP– cartilage oligomeric matrix protein) and the smaller proteoglycans (biglycan, decorin, and fibromodulin). The interaction between highly negatively charged cartilage proteoglycans and type II collagen fibrils is responsible for the compressive and tensile strength of the tissue, which resists applied load *in vivo*.

15.2.2   SYNOVIUM/SYNOVIAL MEMBRANE

Each synovial joint is surrounded by a fibrous, highly vascular capsule/envelope called synovium, which internal surface layer is lined with a synovial membrane. Inside this membrane, type B synoviocytes (fibroblast-like cell lines) are localized/embedded. Their primary function is to continuously extrude high-molar-mass hyaluronans (HAs) into synovial fluid (SF).

15.2.3   SYNOVIAL FLUID

The synovial fluid, which consists of ateultra filtrate of blood plasma and glycoproteins, in normal/healthy joint contains HA macromolecules of molar mass ranging between 6–1nsmega Daltons [7]. SF serves also as a lubricating and shock absorbing boundary layer between moving parts of synovial joints. SF reduces friction and wear and tear of the synovial joint playing thus a vital role in the lubrication and protection of the joint tissues from damage during the motion [8].

The nutrients, including oxygen supply, upon crossing the synovial barrier, permeate through the viscous colloidal SF to the avascular articular cartilage, where they are utilized by the embedded chondrocytes. On the other hand, the chondrocyte catabolites (should) cross the viscous SF prior to being eliminated from the synovial joint [9]. It can thus be concluded that within SF, the process of "mixing" at the joint motion, significantly affects the equilibrium of influx and efflux of all low- and high-molar-mass solutes. It appears that the traffic of solutes is determined by molecular size, with small polar molecules being cleared by venular reabsorption, while high-molecular-sized solutes are removed by lymphatic drainage [10].

15.2.4   HYALURONAN

Figure 3 represents the structural formula of hyaluronan (also called hyaluronic acid, hyaluront) – regularly alternating disaccharide units composed from *N*-acetyl-D-glucosamine and D-glucuronic acid. HA is a polyelectrolyte component of SF; the concentration of HA in healthy human knee SF is 2.5 mg/ml on average [11]. While in the articular cartilage matrix HA is firmly associated via a link protein with proteoglycans (Figure 2), in SF the HA macromolecules are, if at all, only loosely interacting/bound to proteins.

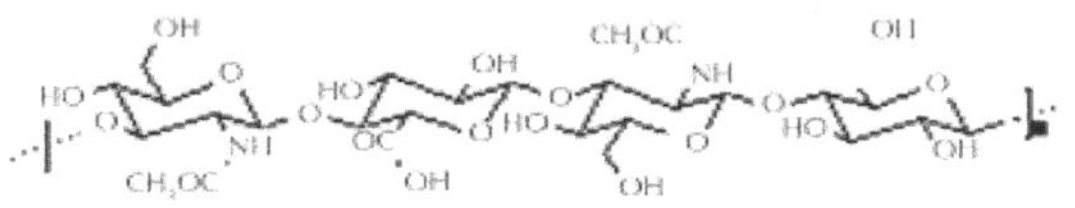

FIGURE 3   Hyalurnn – the acid form.

The HA is a linear non-branched non-sulfated glycosaminoglycan (bio)polymer. In aqueous solutions, HA is represented by negatively charged hyaluronate macromolecules (pKa = 3.21 [12]) with extended conformations, which impart high viscosity/viscoelasticity, accompanied also by low compressibility – the characteristic property of SF [13].

15.2.5   REACTIVE OXYGEN SPECIES IN ARTICULAR CARTIALAR

The articular cartilage is an avascular, acidic (pH 6.6–6.9) and hyperosmotic tissue dependent on diffusion of nutrients supplied mainly from SF (and perhaps partly from subchondral bone [14]) to provide for the metabolic requirements of chondrocytes. The oxygen levels in this tissue are low, ranging between 1 and 6% (cf. Figure 4). While reduction in O2 tension to 6% in all other tissues is already hypoxic, for chondrocytes such oxygen level is normoxic.

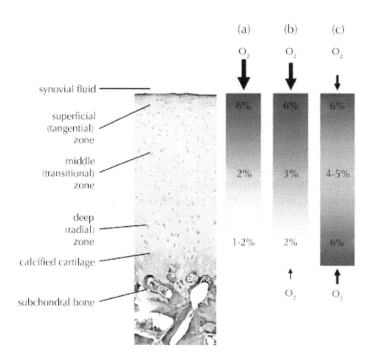

FIGURE 4   The structure of articular cartilage and its oxygen supply (adapted from [14]). Estimated levels of O_2 within the cartilage tissue are shown for three scenarios: (a) Penetration of O_2 exclusively from SF; (b) O_2 supply mostly from SF with a small contribution from subchondral bone; and (c) Supply of O_2 in equivalent amounts from SF and subchondral bone.

In the mitochondria of the eukaryotic cells, not all O_2 is fully reduced to water. A small fraction of oxygen is reduced incompletely yielding reactive oxygen species (ROS), which are assigned to the defense of the organism against viral/bacterial invaders

[15]. It has been established that while ROS content within the articular cartilage tissue remains normal at 6% O_2, it decreased at 1% O_2 [14].

Since hydrogen peroxide generated within the mitochondria of chondrocytes can freely permeate through the chondrocyte cell wall, one should admit the presence of H_2O_2 in all (deep, middle, and superficial) zones of the articular cartilage (Figure 4). The higher the O_2 tension, the greater is the content of H_2O_2 and vice-versa.

The ROS within the cartilage tissue could serve both as intra- and inter-cellular signaling devices and a reactant participating in the so-called Fenton reaction

$$H_2O_2 + Me^{n+} \rightarrow {}^{\bullet}OH + Me^{(n+1)+} + {}^- \qquad (1)$$

where, Me^{n+} and $Me^{(n+1)+}$ represent a (biogenic) transition metal ion in reduced and oxidized state. Among these metals, primarily iron and copper are usually ranked, however, several further trace/biogenic metals can be taken into account as well [1,16].

15.2.6 ROS IN SF AND THEIR FUNCTION THEREOF

The capillaries within synovium continuously provide a plasma filtrate supplying in this way nutrients to the joint tissues (the arterial blood O_2 tension is 13% [17]). This is particularly important for homeostasis of the avascular articular cartilage [10]. As recently stated [16], taking into consideration that articular cartilage does not contain any teleneurons, chondrocytes should perform their autonomic (metabolic) regulation most plausibly using a chemical process, in which both O_2 and ROS play significant roles [17]. To understand this tenet, one should take into consideration that in the joint relaxed state – for example, at night – chondrocytes experience a decreased oxygen supply (a status termed "hypoxia"). However, when the status changes to an enhanced mobility in the morning, joint SF receives elevated supply of O_2 (a situation termed "re-oxygenaion"). Such increased content of oxygen can be, however, deleterious for the homeostasis of the chondrocytes – the cells that in adults lack mitotic activity.

Let us assume that Me^{n+} ions in a given concentration are "entrapped" by (highly) negatively charged cartilage glycosaminoglycans (GAGs) within the superficial (tangential) zone of the articular cartilage (Figure 4). During the utilization of O_2 – respirain – by chondrocytes, a limited amount of H_2O_2 liberated from their mitochondria can react with the entrapped transition metal ions generating hydroxyl (${}^{\bullet}OH$) radicals. Due to extremely short half-life of these species (picoseconds), they react *in situ nascendi* with GAGs – chondroitin sulfate (CS) and/or keratan sulfate (KS). The C-type radicals of CS or KS can, however, instantly undergo a reaction of hydrogen radical transfer onto the neighboring HA macromolecules within the SF. In such a way, free C-(mcals)radicals of hyaluronan appear nearby the superficial zone of the articular cartilage. And it is this very C-(macro)radical (denoted later as $A^{\bullet}$), which reacts and in this way reduces the (free "hyperoxic") O_2 tension within and nearby the superficial zone of the articular cartilage – according to the reaction presented in the following scheme:

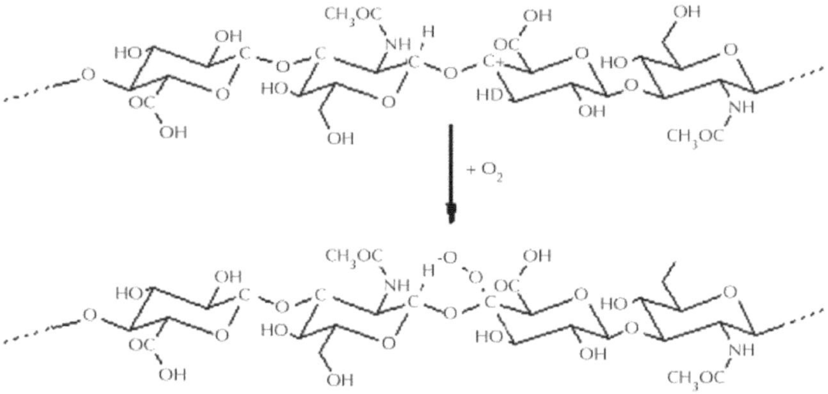

SCHEME 1   Entrapment of oxygen by the hyaluronan C-(macro)radical (A·) yielding a peroxyl (macro)radical (A–O-O·).

or briefly

$$A–H + ·OH \rightarrow A· + H_2 \tag{2}$$

$$A· + O_2 \rightarrow A–O–O \tag{3}$$

where, A–H represents the intact hyaluronan macromolecule (Fig.Figure 3 and Scheme 1).

Subsequently, this A–O-O· peroxyl (macro)radical can transform simply by an in-tramolecular 1,5-hydrogen shift to another C-(macro)radical – A· (cf. Scheme 2). By participation of another O$_2$ molecule, this A· radical can yield two fragments of the HA biopolymer: (i) the fragment, which possesses an aldehyde terminus, and (ii) the fragment bearing a hydroperoxide functional group. It is naturally evident that both fragments differ in their chemical structure from the initial HA macromolecule, not only due to the included novel substituents (–C=O; –O-OH) but above all by a reduced molar mass of both polymer fragments compared to that of the parent biopolymer.

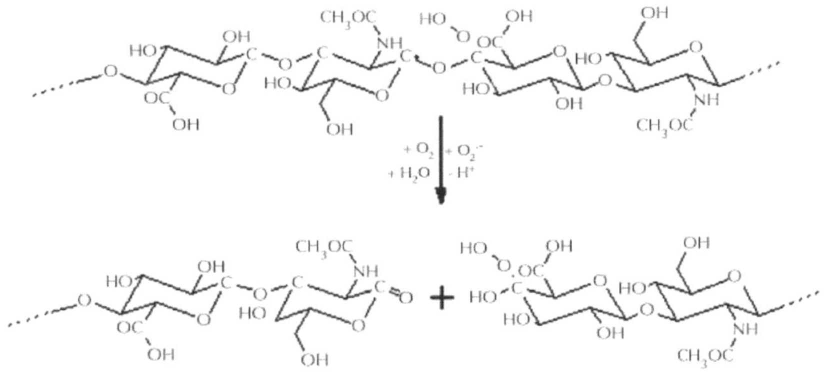

SCHEME 2   Strand scission of the C-(macro)radical (A˙) yielding two fragments.

Since the intermolecular reaction between the CS and KS radicals and the native HA macromolecule could yield various A˙ radicals – formed for example at C(4) of the D-glucuronate/D-glucuronic acid (GlcA) unit (cf. Scheme 1) or at C(1) of GlcA unit, as well as at C(1) or C(3) of N-acetyl-D-glucosamine (GlcNAc) [18] – various biopolymer fragments are produced.

Very recently Kennett and Davies [19] reported the data obtained with both the C(1)- and the C(2)- ^{13}C-labeled N-acetyl-D-glucosamine, and the apparent highly selective generation of radicals at the C(2) position of the isopropyl group of the β-isopropyl glycoside, which allow the authors to rationalize the specific banding pattern observed on oxidation of hyaluronan: The lack of reactivity at C(1)/C(2) of the N-acetyl-D-glucosamine monomers and the specific formation of radicals on the isopropyl group, which models the C(4) glycosidic linkage site of the glucuronic acid, implicate attack at C(4) of the glucuronic acid subunits and subsequent β-scission of this radical as a major route to cleavage of the hyaluronan backbone (Scheme 3). A contribution from reaction at C(1) of the glucuronic acid and subsequent cleavage of the alternative glycosidic linkage cannot be discounted; however, it is clear that an alternative route involving C(3) on the N-acetyl-D-glucosamine monomer is less favored, as only low levels of initial hydrogen atom abstraction seem to occur at this position as judged by the low yield of radicals that did not have additional ^{13}C couplings observed with the two labeled N-acetyl-D-glucosamine species. It should be pointed, however, that the products of the hyaluronan strand cleavage depicted in Scheme 3 do not take into account that the ubiquitous oxygen participate within the strand scission reaction and thus, analogously to Scheme 2, the involved O_2 molecule with the A˙ radical yields two fragments of the HA biopolymer: (i) the fragment bearing a hydroperoxide functional group, and (ii) the fragment, which possesses an aldehyde terminus. As stated above, both fragments naturally differ in their chemical structure due to the included –C=O or –O-OH substituent and, above all, by the reduced molar mass of both polymer fragments compared to that of the parent HA biopolymer.

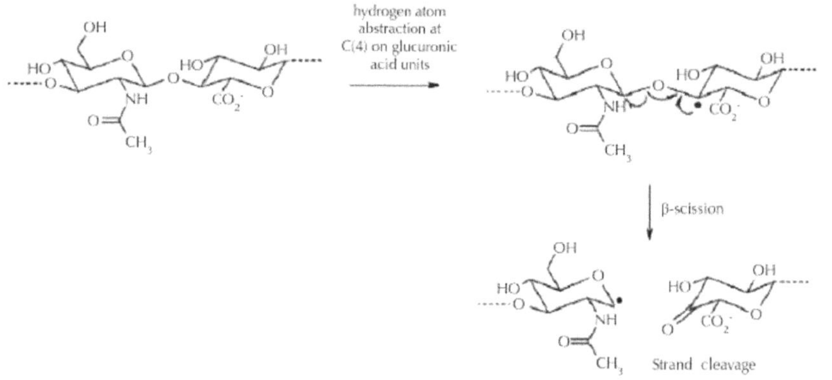

SCHEME 3   Potential mechanism of hyaluronan strand cleavage as a result of hydrogen abstraction and radical formation on C(4) of the glucuronic acid unit (adapted from [19]).

Along with the fragmentation reactions shown in Schemes 2 and 3, the radical attack on the GlcA and GlcNAc moieties can also lead to the ring opening without breaking the polymer chain [11,18,20,21].

There exists, however, a remarkable phenomenon of *in vivo* free-radical oxidative degradation of hyaluronan: Under physiological conditions, the SF viscosity does not undergo any changes since the content of "naive" hyaluronan remains constant due to permanent *de novo* production of mega Dalton HA macromolecules by (stimulated) type B synoviocytes. Thus, the self-perpetuating oxidative (non-enzymatic)HA "catabolism" in SF represents a rather delicate and properly balanced mechanism that presumably plays significant role in regulating the physiological – normoxygen – homeostasis for chondrocytes. At the same time, the produced polymer fragments, which are probably cleared from the joint by drainage pathways, serve most likely as chemical messengers/feedback molecules. These play role in the adjustment of the optimum mode of functioning of the synovial membrane and of the HA-producing cells, B synoviocytes, localized within. In other words, during physiologic joint functioning, the hyaluronan in SF plays the role of a "scavenger antioxidant", whereas the produced polymer fragments can subsequently serve as messengers mediating information on the changes occurring in the homeostasis of the joint [16].

High "protective/scavenging efficiency" of hyaluronan against the *in vitro* action of ·OH radicals has been earlier pointed out by some authors [22,23]. Presti and Scott [23] described that high-molar-mass hyaluronan (megaDalton HA) was much more effective than the lower-molar-mass HAs (hundreds of kiloDaltons HAs) in scavenging ·OH radicals generated by a Fenton-type system comprising glucose and glucose oxidase *plus* Fe^{2+}-EDTA chelate.

15.2.7   HYPOXIA AND RE-OXYGENATION OF THE JOINT

As SF of healthy human exhibits no activity of the hyaluronidase enzyme, it has been inferred that oxygen-derived free radicals are involved in a self-perpetuating process of HA catabolism within the joint [24]. This radical-mediated process is considered to account for ca. twelve-hour half-life of native HA macromolecules in F.

To understand how to maintain a radical reaction active/self-perpetuating, its propagation stage should first be analyzed. If a peroxyl-type (macro)radical (A–O-O·) exists within SF, due to the relatively high reactivity of the unpaired electron on oxygen, the following intermolecular reaction can be assumed

$$A\text{–}O\text{-}O^{\cdot} + A\text{–}H \rightarrow A\text{–}O\text{-}OH + \qquad\qquad (4)$$

In the case when A· is a C-type (macro)radical, it is this very reactant that traps the dioxygen molecule, dissolved in SF, according to the reaction

$$A^{\cdot} + O_2 \rightarrow A\text{–}O\text{-}O^{\cdot} \qquad\qquad (5)$$

Hence, by combining the reactions 4 and 5, the net reacion,

$$A\text{–}H + O_2 \rightarrow A\text{–}O\text{-} \qquad\qquad \text{(net reaction)}$$

corroborates the statement that one particular function of (a high-molar-mass) HA is to trap the oxygen excess during the phase of joint re-oxygenation [16].

15.3 PHYSIOLOGIC OXIDATIVE CATABOLISM OF HYALURONAN: PARTICIPATION OF BIOGENIC TRANSITION METAL IONS

As stated in Scheme 2 and reaction 4, A–O-OH hydroperoxides are generated during the self-perpetuating – propagation – stage of the hyaluronan oxidative catabolism. The fate of A–O-OH type hydroperoxides, however, is significantly dependent on the presence or absence of the transition metal ions within SF. In the former case, the following reactions could be suggested for decomposition of the generated A–O-OH hydroperoxides

$$A\text{–}O\text{-}OH + Me^{n+} \rightarrow A\text{–}O^{\cdot} + HO^- + Me^{(n+1)} \qquad\qquad (6)$$

$$A\text{–}O\text{-}OH + Me^{(n+1)+} \rightarrow A\text{–}O\text{-}O^{\cdot} + Me^{n+} + {}^+ \qquad\qquad (7)$$

As can be seen, while the "propagator" that participates in reaction 4 is (re)generated by reaction 7, reaction 6 produces an alkoxyl type (macro)radical A–O·. The ratio of the A–O-O· to A–O· radicals is, however, governed by the present transition metal ions, or, more precisely, by the ratio of $Me^{(n+1)+}$ to Me^{n+}. To answer the question, which transition metals may be present in SF and cells or tissues of healthy human beings, one should take into account the data presented in Tables 1 and 2.

TABLE 1   Contents of transition metals in blood serum of healthy human volunteers and in *post mortem* collected SF from subjects without evidence of connective tissue disease

Element	Mean concentration in blood serum [µg/100 mL][a]	Mean concentration in synovial fluid [µg/100 g][a]
Iron	131.7 (23.6)[b]	29.0 (5.19)[b]
Copper	97.0 (15.3)	27.5 (4.33)
Zinc	115.4 (17.7)	17.6 (2.69)
Manganese	2.4 (0.44)	2.4 (0.44)
Nickel	4.1 (0.70)	1.2 (0.20)
Molybdenum	3.4 (0.35)	1.0 (0.10)

[a]Reported by Niedermeier and Griggs [25].
[b]Data in parentheses are the values in µM calculated in assumption that 100 g of SF has a volume of 100 mL.

TABLE 2   Average relative abundance of some biogenic transition metals in the mammalian blood plasma and cells/tissues

Element	Blood plasma [µM][a]	Cell/Tissue [µM][a]
Iron	22	≈ 68
Copper	8-24	0.001-10
Zinc	17	180
Manganese	0.1	180
Nickel	0.04	2
Molybdenum	-	0.005

[a]Adapted from [26].

Based on the data listed in Table 1, iron and copper are the two prevailing redox active transition metals in SF. It should be, however, pointed out that the respective concentrations of ca. 5.2 µM of iron ions and 4.3 µM of copper ones do not represent those, which are (freely) disposable to catalyze the oxidative catabolism of hyaluronan within SF. As has been reported, the availability of iron to stimulate *in vivo* generation of ·OH radicals is very limited, since concentrations of "free" iron, are seldom larger than 3 µM in human samples [27].

Let us now deal with the oxidation states of iron within SF of a healthy human. By accepting that the concentration of ascorbate in SF of healthy subjects reaches the values close to those established in blood serum, that is 40–140 µM [28], it must be admitted that the transition metal ions in SF of a healthy human being are in the reduced oxidation state, that is Me^{n+}. Thus, in the case of the ascorbate level, which many times exceeds the concentration of transition metal ions, the actual concentration of ferrous ions should exceed that of ferric ones, and thus A–O· radicals should prevail. These radicals could, similarly to the A–O-O· ones, propagate the radical chain reaction as follows

$$A\text{–}O^{\boldsymbol{\cdot}} + A\text{–}H \rightarrow A\text{–}OH + A^{\boldsymbol{\cdot}} \qquad (8)$$

Yet, due to the redox potential of the pair RO·,H⁺/ROH = + 1.6 V, which surpasses significantly that of ROO·,H⁺/ROOH = + 1.0 V, the actual content of A–O· in SF is practically nil; the half-life of the A–O· radicals is much shorter than that of A-O-O· ones – microseconds *vs.* seconds.

15.3.1 OXIDATIVE/NITROSATIVE STRESS

Oxidative and/or nitrosative stress are terms used to describe situations, in which the organism's production of oxidants exceeds the capacity to neutralize them. The excess of oxidative species can cause "fatal" damage to lipids within the cell membranes, cellular proteins and nucleic acids, as well as to the constituents of the extracellular matrix, such as collagens, proteoglycans, and so on. [29].

Oxidative and/or nitrosative stress has been implicated in various pathological conditions involving several diseases, which fall into two groups:

1. Diseases characterized by "inflammatory oxidative conditions" and enhanced activity of either NAD(P)H oxidase (leading to atherosclerosis and chronic inflammation) or xanthine oxidase-induced formation of oxidants (implicated in ischemia and reperfusion injury),
2. Diseases characterized by the implication of pro-oxidants that shift the thiol/disulphide redox equilibrium and cause impairment of glucose tolerance - the so-called "mitochondrial oxidative stress" conditions (leading to cancer and diabetes mellitus) [3].

15.3.2 OXIDANTS

In a broader sense, oxidation concerns the reaction of any substance with molecules of oxygen, the primary oxidant. In chemistry, however, the term "oxidant" is used for all species able to render one or more (unpaired) electrons.

TABLE 3 Main ROS and RNS

Radical		Non-radical	
hydroxyl	·OH	peroxynitrite anion	ONOO⁻
superoxide anion radical	$O_2^{\cdot -}$	hypochloric acid	HOCl
nitric oxide	·NO	hydrogene peroxide	H_2O_2
thyil	–RS·	singlet oxygen	$^1\Delta_g \, (^{-1}O_2)$
alkoxyl	RO·	ozone	O_3
peroxyl	ROO·	nitrosyl cation	NO⁺
		nitroxyl anion	NO⁻
		nitryl chloride	NO_2Cl

In a simplified way, oxidants can be classified as free-radical and non-radical species (cf. Table 3; adapted from [30]). They are often classified as reactive oxygen species (ROS) and reactive nitrogen species (RNS). Although the latter, similarly to ROS, contain oxygen atom(s) – for example, NO^+, NO^-, and NO_2Cl – the RNS usually participate at nitrosylation reactions.

15.3.3 OXYGEN METABOLISM–SOURCE OF ENERGY

Several oxidant species are produced at the processes occurring in animal cells, including human ones, during metabolism of oxygen, when these cells generate energy. Although the substrate (O_2) is – by a cascade of enzymatically driven reactions – reduced within subcellular organelles, mitochondria, to a completely harmless substance, the waste product – water, a fraction of generated ROS may escape from the enzymatically controlled processes:

$$O_2 + 1e^- \rightarrow O_2^{\cdot-} \tag{9}$$

$$O_2^{\cdot-} + 1e^- + 2H^+ \rightarrow H_2O_2 \tag{10}$$

$$H_2O_2 + 1e^- + H^+ \rightarrow {\cdot}OH + H_2O \tag{11}$$

$${\cdot}OH + 1e^- + H^+ \rightarrow H_2O \tag{12}$$

net reaction

$$O_2 + 4e^- + 4H^+ \rightarrow 2H_2O \tag{13}$$

As indicated by the reaction steps (9), (10), and (11), oxidants, namely $O_2^{\cdot-}$, H_2O_2, and ${\cdot}OH$ are intermediate products of the enzymatically controlled cascade. Their reactivity and presumable site of action can be assessed by physico-chemical parameters, such as standard reduction potential (E^0) and half-life ($t_{1/2}$) of the given species (cf. Table 4).

TABLE 4   Standard reduction potential (and half-life) for some dioxygen species in water, pH 7, 25°C[a]

Species	(reaction)	E^0 [V]	$t_{1/2}$ [s]
O_2	(9)	-0.33[b]	reactive
$O_2^{\cdot-}$	(10)	$+0.89$	10^{-6}
H_2O_2	(11)	$+0.38$	long living
${\cdot}OH$	(12)	$+2.31$	10^{-9}

[a]Adapted from [31].
[b]The greater the positive E^0 value, the greater is generally the species reactivity, that is the ability to catch an electron [cf. reactions (9)–(12)].

With regard to the high (positive) value of E^0 and to the short half-life values, escape of $\cdot OH$ and $O_2{}^-$ from the sphere immediately surrounding mitochondrion can be virtually excluded. Yet the neutral molecule H_2O_2 is considered to be movable one, which can escape as from the "body" of the mitochondrion as well as from the cell body itself. It is comprehensible that in some tissues the actual H_2O_2 concentrations may reach 100 μM or more as for example, in human and other animal aqueous and vitreous humors. The hydroperoxide levels at or below 20–50 μM seem, however, to have limited cytotoxicity to many cell types [32].

15.4 OXYGEN METABOLISM–A DEFENSE MECHANISM AGAINST VIRAL/ BACTERIAL INVADERS

Along with the above four-electron reaction (13), several specialized cells – or more precisely their specific (sub)cellular structures – are able to reduce O_2 molecules producing the superoxide anion radical, which in aqueous (acidic) milieu can form the reactive perhydroxyl radical ($\cdot O_2H$).

Nitric oxide, called also nitrogen monoxide ($\cdot NO$), a (bioactive) free radical, is produced in various cells/tissues by NO-synthase (NOS) enzymes. The three distinct NOS isoforms are P_{450}-related hemoproteins that during L-arginine oxidation to L-citrulline produce $\cdot NO$. Two of the permanently present enzymes that participate in the regulation of the blood vessel tonus are termed constitutive NOS (cNOS), while the third one is called an inducible NOS (iNOS). The level of $\cdot NO$ produced by iNOS increases markedly during inflammation, a process accompanied with abundant production of the superoxide anion radical.

The two radical intermediates – $O_2{}^-/\cdot O_2H$ and $\cdot NO$ – serve as precursors of various ROS and RNS, including hydrogen peroxide, peroxynitrite/peroxynitrous acid, hypochlorous acid, and so on. On respiring air, human beings by utilizing one mole of O_2 ingest 6.023×10^{23} molecules of oxygen, of which approximately 1–3% is assigned to the generation of ROS/RNS that defend the organism against viral/bacterial invaders [15].

It has been noted that certain organ systems are predisposed to greater levels of oxidative stress and/or nitrosative–stress. Those organ systems most susceptible to damage are the pulmonary system (exposed to high levels of oxygen), brain (exhibits intense metabolic activity), eye (constantly exposed to damaging UV light), circulatory system (victim to fluctuating oxygen and nitric oxide levels) and the reproductive systems (at risk from the intense metabolic activity of sperm cells) [30]. In some cases, however, the intermediate and/or the "final" reactive oxidative species may also damage cells/tissues of the human host. Imbalance between the extent of damage and self-repair of the functionally essential structures may result in a broader host tissue injury, eventually leading to a specific disease.

Because of the highly reactive nature of ROS/RNS, it is difficult to directly demonstrate their presence *in vivo*. It is considerably more practical to measure the "footprints" of ROS and RNS, such as their effects on various lipids, proteins, and nucleic acids [29].

15.4.1 INDIRECT ROS/RNS EVIDENCE

Most ROS/RNS have very short half-live times thus they cannot be directly detected in the organisms. That is why, as reported also by Valko *et al.* [3], convincing evidence for the association of oxidative/nitrosative stress and acute and chronic diseases lies on validated biomarkers of these stresses. Table 5 summarizes most representative biomarkers of oxidative damage associated with several human diseases.

TABLE 5 Biomarkers of oxidative damage associated with several chronic human diseases (adapted from [3])

Disease Biomarker[a]	Alzheimer's disease	Atherosclerosis	Cancer	Cardiovascular disease	Diabetes mellitus	Parkinson's disease	Rheumatoid arthritis
8-OH-dG			+				
Acrolein		+		+			
AGE	+				+		
Carbonylated proteins						+	
F_2-isoprostanes	+	+		+	+		+
GSH/GSSG	+		+	+	+	+	+
HNE	+	+		+		+	
Iron level						+	
MDA	+	+	+		+		
NO_2-Tyr	+	+	+	+	+		
S-glutathiolated proteins					+		

[a]Abbreviations: *8-OH-dG, 8-hydroxy-20-deoxyguanosine; AGE, advanced glycation end products; GSH/GSSG, ratio of glutathione/oxidized glutathione; HNE, 4-hydroxy-2-nonenal; MDA, malondialdehyde; NO_2-Tyr, 3-nitro-tyrosine*

There are numerous further diseases whose pathology involves reactive oxidative/ oxygen-derived species that is ROS and/or RNS, at the onset and/or at later stages of the disease [33]. The magnitude and duration of the change in the concentrations of these species appear to belong among the main regulatory events (cf. Figure 5).

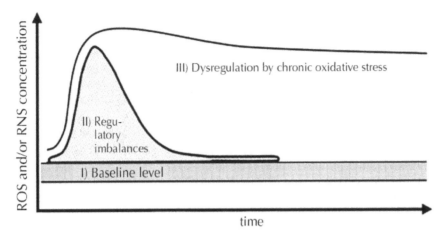

FIGURE 5   Regulatory events and their dysregulation depend on the magnitude and duration of the change in ROS and/or RNS concentration(s) (adapted from [34]).

Today it is a widely accepted fact that ROS and RNS normally occur in living tissues at relatively low steady-state levels (cf. Figure 5, stage I "Baseline level"). The regulated increase in the production of superoxide anion radical or nitric oxide leads to a temporary imbalance, which forms the basis of redox regulation (stage II in Figure 5, "Regulatory imbalances"). The persistent production of abnormally large amounts of ROS or RNS, however, may lead to persistent changes in signal transduction and gene expression, which, in turn, may give rise to pathological conditions (as seen in Figure 5, stage III "Dysregulation by chronic oxidative stress") [34]. One of the classes of such diseases includes arthritic conditions – inflammatory diseases of joints. A substantial amount of evidence exists for an increased generation of oxidants in patients suffering from acute and chronic inflammatory joint diseases [36,37] – see Table 6.

TABLE 6   Some characteristics registered within SF during inflammatory joint diseases[a]

Diagno-sis	SF viscosity	Blood characteristics			
		White cells/ μL	% of PMNLs	PMNLs/ μL	H_2O_2 flux [μM/min]
Healthy	normal	<200	7	<14	<0.003
OA	decreased	600	13	48	0.017
RA	decreased	1900	66	1254	0.276

[a]Adapted from [37].

15.4.2   *REGULATORY IMBALANCES WITHIN A SYNOVIAL JOINT*

As schematically reported by Dröge [34], under physiological status, "Baseline level" (cf. Figure 5) of ROS and/or RNS concentration play an important role as regulatory mediators in signaling processes. In case of the composition of SF of healthy organisms, one may state two border concentrations of ROS (and RNS as well), which are primarily determined by the O_2 level within SF, or more precisely by the H_2O_2 level escaped from mitochondria of chondrocytes and from those of cells of the synovial membrane. A lower one exists at rest regimen of the joint and a higher H_2O_2 level at reoxygenation of the joint tissues during movement of the subject. The high-molar-mass HA however keeps most probably the joint ROS/RNS homeostasis between the two concentration values inside the "Baseline level" (see Figure 5, stage I).

On accepting the tenet that concentrations of H_2O_2 ranging around 50 µM (sometimes even up to 100 µM) are not toxic to any cells [32], the highest limit (cf. stage I, Figure 5) of the hydrogen peroxide level in SF, and thus in contact with both chondrocytes and synovial-membrane cells, is close to this concentration (<100 µM). The flux of H_2O_2 in the amount of less than 0.003 µM per minute does not change SF viscosity (cf. Table 6). In light of this observation one can propose that the ROS action, that is H_2O_2-degradative action on the high-molar-mass HA, is fully compensated by the *de novo* synthesis of megaDalton hyaluronans by the synoviocytes embedded within the synovial membrane of healthy human beings. Our detailed studies focusing on the H_2O_2-degradative action to HA macromolecules also showed that hydrogen peroxide up to hundreds of micromolar concentrations led to practically no cleavage/decay of high-molar-mass hyaluronan samples when the reaction system was "free" of any transition metal ions, namely those of iron and/or copper [M. Stankovská *et al.*, not published].

Let us now admit the situation of occurrence of temporary "Regulatory imbalances" (stage II in Figure 5), or more precisely the situation at which an acute inflammation is initiated within the synovial joint. On taking into account the data given in Table 6, the increase in ROS concentration, or more precisely the increase in H_2O_2 flux, appears to be functionally related to the rising number of PMNLs in the SF, presenting in the initial phase as Regulatory inbalance. This increase is however associated with the following events: i) infiltration of the increased number of white cells (PMNLS and/or macrophages) from the blood circulation into the SF, and ii) activation of these cells in the SF. Yet concerning the event given in ii), it has to be emphasized that at the time of infiltration movement of the white blood cells is impeded in the SF, due to its viscosity, which can be characterized as "normal" (cf. Table 6; see Figure 6) or high caused by the presence of high-molar-mass HA macromolecules. Moreover, it is a well-known fact that especially high-molar-mass hyaluronans exert antiimflammatory action or more precisely, the long-sized HA chains quench the PMNLs and macrophages.

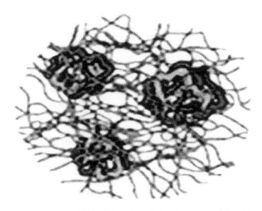

FIGURE 6   The movement of the white blood cells in the normal/highly viscous SF. The long-sized HA chains are sketched as blue strands.

Thus one may admit that infiltration of an increased number of white cells into a millieu such as that of SF of healthy human beings need not immediately result in a rise of the ROS concentration or the H_2O_2 level enhancement, respectively. The demand of rapid/acute growth of ROS/RNS level within the joint during the stage II (cf. Figure 5, "Regulatory imbalances") could not be met in this way. Resulting from our experimental findings, we may hereby offer/recommend our hypothesis/speculation in point of process sequencing which can very quickly, owing to their physiological status, bring about – for a temporary time period – the status possibly be defined as accute inflammation, or – by taking into account the Dröge scheme (cf. Figure 5 [34]) – the "Regulatory inbalances".

INFLAMMATION

Inflammation generally means a complex biological response of tissues to harmful stimuli, such as infective pathogens, damaged cells, toxins, physical and/or chemical irritants. It is a protective attempt by the organism to remove injurious stimuli and to initiate the healing process for the tissue. Yet inflammation that runs unchecked can lead to various diseases (cf. Table 5), including those connected to synovial joints. Normally, however, inflammation is critically controlled and closely regulated by the body.

Inflammation can be classified as acute or chronic (Table 7). Acute inflammation is the initial response of the body to harmful stimuli and is achieved by the increased movement of PMNLs from the blood into the injured tissues. Then a cascade of biochemical events propagates and matures the (local) inflammatory response. Chronic inflammation usually leads to a progressive shift in the type of immune cells which are present at the site of inflammation and is characterized by destruction and often by (partial) healing of damaged tissues.

TABLE 7   Comparison between acute and chronic inflammation (from [38])

Inflammation	Acute	Chronic
Causative agent	Pathogens, injured tissues	Persistent acute inflammation due to non-degradable pathogens, persistent foreign bodies, or autoimmune reactions
Major cells involved	Neutrophils, mononuclear cells (monocytes, macrophages)	Mononuclear cells (monocytes, macrophages, lymphocytes, plasma cells), fibroblasts
Primary mediators	Vasoactive amines, eicosanoids	IFN-γ and other cytokines, growth factors, reactive oxygen species, hydrolytic enzymes
onset	immediate	delayed
duration	few days	up to many months or yr
outcomes	resolution, abscess formation, chronic inflammation	tissue destruction, fibrosis

Acute inflammation – a short-term process appearing in a few minutes or hours – is usually characterized by five cardinal signs: rubor, calor, tumor, dolor, and *functio laesa*. However, the acute inflammation of an internal organ may not be manifested by the full set of signs.

Inflammation, and especially the acute one, is associated with elevated systemic levels of acute-phase proteins. These proteins prove beneficial in acute inflammation.

ACUTE-PHASE PROTEINS

The acute-phase proteins are a class of proteins whose plasma concentrations increase (positive acute-phase proteins) or decrease (negative acute-phase proteins) in response to inflammation. This response is called the acute-phase reaction or acute-phase response. The acute-phase reactants are produced by the liver in response to specific stimulations. The following positive acute-phase proteins belong to the physiologically most prominent ones: C-reactive protein, α_1-antitrypsin and α_1-antichymotrypsin, fibrinogen, prothrombin, complement factors, ferritin, serum amyloid A, α_1-acid glycoprotein, ceruloplasmin, and haptoglobin. Others – negative acute-phase proteins such as albumin, transferrin – give negative feedback on the inflammatory response.

CERULOPLASMIN

The concentration of ceruloplasmin, whose molar mass ($\approx$ 134 kDa) exceeds nearly twice that of albumin, increases markedly under certain circumstances – including those of acute inflammation. Since each ceruloplasmin macromolecule complexes/ binds up to eight Cu(II)/Cu(I) ions of which two can liberate relatively easily [39], at the early stage of acute inflammation the actual copper level increases markedly. The consequence of higher ceruloplasmin concentration in blood plasma – accompanied with a rise in the concentration of copper ions – would mean a larger amount of this biogenic trace element that might cross the synovial membrane [16]. Yet, due to the gel-like consistency of SF, the copper ions entering into this specific environment start their redox action in the vicinity of the synovial membrane.

15.4.3   WEISSBERGER'S OXIDATIVE SYSTE

The concentration of ascorbate in SF of healthy subjects reaches the values close to those established in blood seru,.that is 40–140 µM [28]. Ascorbate, an "actor of physiologic HA catabolism in SF" with copper liberated from ceruloplasmin, creates easily the so-called Weissberger's oxidative system [40,41] – ascorbate-Cu(I)-oxygen – generating H_2O_2 (cf. Scheme 4) [42-44]. Moreover, due to the simultaneous decomposition of hydrogen peroxide by the redox active copper ions, a large flux of hydroxyl radicals may occur [45].

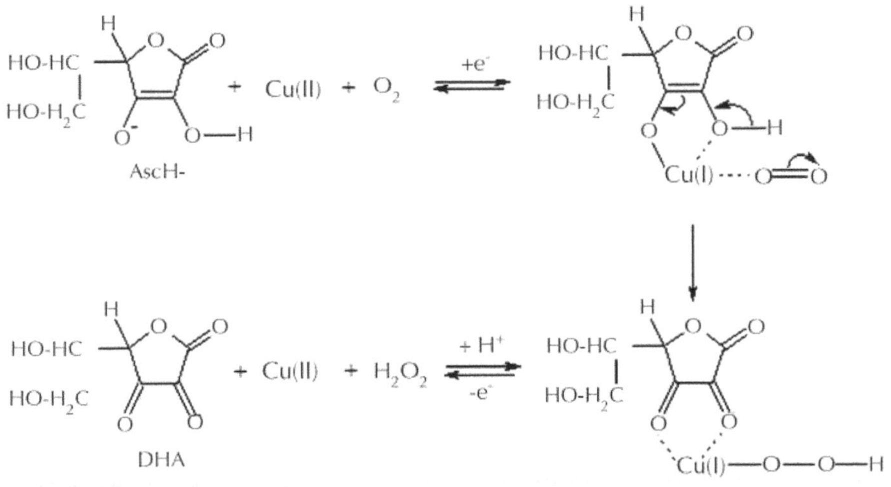

SCHEME 4   Generation of H_2O_2 by Weissberger's system from ascorbate and Cu(II) under aerobic conditions (adapted from Fisher and Naughton [44]).

As evident from the data listed in Table 1, iron and copper are the two prevailing redox active transition metals in SF. Although just only a minor fraction of their respective total levels equaling 5.2 µM and 4.3 µM is disposable for Weissberger's and/ or Fenton-type reactions, it are the copper ions that better fulfill the requirement of acute (rapid) generation of ROS – particularly of ˙OH radicals (cf. Figure 7).

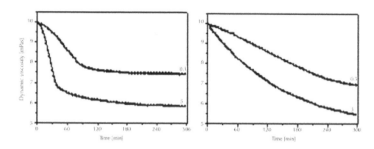

FIGURE 7   Time dependences of dynamic viscosity of solutions of a high-molar-mass HA sample.

- *Left panel*: Solutions of the HA sample with addition of 100 µM ascorbic acid immediately followed by admixing 0.1 or 5 µM of $CuCl_2$.
- *Right panel*: Solutions of the HA sample with addition of 100 µM ascorbic acid immediately followed by admixing 0.5 or 5 µM of $FeCl_2$.

Figure 7 illustrates the degradative action of ROS by monitoring the viscosity-time profiles of a HA solution into which – along with 100 µM ascorbate – a single transition metal was added [46]. As evident, a significant reduction of the solution dynamic viscosity (η), corresponding to the degradation of the high-molar-mass HA sample, clearly indicates a concentration-dependent manner for each metal (cf. left and right panels ig.Figure 7). While the character of the time dependence of η value upon the addition of $FeCl_2$ (5.0 µM) can be described as a gradual monotonous decline, the addition of $CuCl_2$ (5.0 µM) resulted in a literally "dramatic" drop of η value within a short time interval (30 min). A similar drop of η value and two-phase reaction kinetics are identifiable upon the addition of even a minute (0.1 µM) amount of $CuCl_2$ (seg. Figure 7, left panel). A possible explanation of this dissimilarity lies most probably in different reaction kinetics of the processes leading to generation of oxygen-derived reactive species in the system ascorbate *plus* $CuCl_2$ and in that comprising ascorbate *plus* $FeCl_2$.

As seen in Figure 7, the transition metal – either iron or copper – can play an active role in oxidative HA catabolism. However, the increase in Cu(II) concentration within the joint (and particularly in SF) could lead to an extremely rapid degradation of the native HA macromolecules. How efficiently the chemically generated ˙OH radicals are "scavenged" within this microenvironment by the locally disposable albumin as well as by the HA polymer fragments of lower molecular size, remains questionable. The oxidative process may escape the control mechanisms and damage/disrupt the synovial membrane. Moreover, the intermediate-sized HA-polymer fragments generated within this microenvironment could participate in the activation of "defender" cells. They may further intensify the inflammation state of the injured tissue(s) as the HA-polymer fragments can in turn augment the inflammatory responses. As reported by Jiang *et al.*, the HA fragments in the, for example, 2×10^5 Da range induce the expression of a number of inflammatory mediators in macrophages, including chemokines, cytokines, growth factors, proteases, and nitric oxide [47]. In this way, the oxidants generated by activated defender cells may enlarge the damage within the involved joint tissues such as the synovial membrane (cf. Figure 8). Such an increase in unmediated reactive radicals, generally termed oxidative stress, is an active area of research in a variety of diseases where copper may play an insidious role.

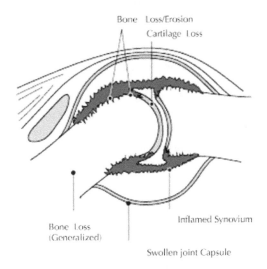

FIGURE 8   Damages within the inflamed joint tissues.

Moreover, reactive oxygen species appear to disrupt copper binding to ceruloplasmin, thereby releasing "free" copper ions, which in turn may promote oxidative pathology [39]. The damage can be manifested by visually localizable cardinal signs of inflammation – that is rubor, calor, tumor, dolor, and *functio laesa*, yet less distinct, repeated (micro-acute) inflammatory injures may lead to a disastrous outcomeg.for example, an autoimmune disease such as rheumatoid arthritis.

15.5   RELEVANCY AND FUNCTION OF WEISSBERGER'S OXIDATIVE SYSTEM AT ACUTE INFLAMMATION OF THE JOINT

As demonstrated by the results depicted in Figure 7 (left panel) Weissberger's oxidative system is really a prompt/ultimate generator of hydrogen peroxide leading immediately to dramatic flux of •OH radicals. Subsequently these radicals initiate a significant degradation of long-chain HA macromolecules, the process which diminishes markedly the dynamic viscosity of the hyaluronan solution. A similar HA degradative process can be anticipated in SF at the early stage of acute (synovial) joint inflammation. The lower SF viscosity may markedly promote the transition of defender cells from blood through the synovial membrane and further enhance the movement of these cells to the target synovial and periarticular tissues. These cells may simultaneously undergo activation in contact with/binding to biopolymer fragments resulted from (˙OH) radical degradation of native high-molar-mass hyaluronans present in SF. The infiltrated defender cells thus may start their more or less specific action inside the intraarticular space.

15.5.1 CHRONIC INFLAMMATION

In acute inflammation, if the injurious agent persists, chronic inflammation will ensue. This process marked by inflammation lasting many days, months or eversyears, may lead to the formation of a chronic wound. Chronic inflammation is characterized by the dominating presence of macrophages in the injured tissue. These cells are powerful defensive agents of the body, but the "toxins" they release – including ROS and/or RNS – are injurious to the organism's own tissues. Consequently, chronic inflammation is almost always accompanied by tissue destruction. Destructed tissues are recognized by the immunity system and, when "classified" by the body as foreign ones, a cascade of autoimmune reactions could start. Such reactions are well established in diseases such as rheumatoid arthritis, where – along with the (synovial) joints – several further tissues/organs e.g. for example, lungs, heart, and blood vessels, are permanently atacked e.g. that is misrecognized as foreign ones.

15.5.2 MEDICATIONS USED TO TREAT INFLAMMATORY JOINT DISEASES

There are many medications available to decrease joint pain, swelling, inflammation and to prevent or minimize the progression of the inflammatory disease. These medications include:

- Non-steroidal anti-inflammatory drugs (NSAIDs – such as acetylsalicylic acid/ aspirin, ibuprofen or naproxen).
- Corticosteroids (such as prednisone).
- Anti-malarial medications (such as hydroxychloroquine).
- Other medications, including methotrexate, sulfasalazine, leflunomide, anti-TNF medications, cyclophosphamide, and mycophenolate.

As reported in the Section "Relevancy and function of Weissberger's oxidative system at acute inflammation of the joint", the early acute-phase of (synovial) joint inflammation should, most plausibly, be accompanied with generation of ROS (and RNS) – particularly with ˙OH radicals. These, however, due to their extremmly high electronegativity (-2.31 V) should – in contact with any hydrogen atom containing compounds – entrap a proton (˙H). By that process the ˙OH radicals are partially or fully scavenged (cf. Figure 9). If the resulting radical generated from the given compound/medication is not able to initiate HA degradation, we speak of drug-scavenging, which could moderate the free radical process within the inflamed joint.

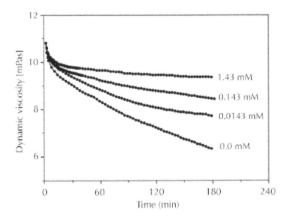

FIGURE 9   Effect of acetylsalicylic acid on HA degradation in the system 0.1 μM CuCl₂ + 100 μM ascorbic acid + 2 mM NaOCl.

Concentration of acetylsalicylic acid added into the system before initiation of HA degradation in mM: 0.0, 0.0143, 0.143, and 1.43.

Figure 9 illustrates such an *in vitro* testing of the scavenging efficiency of acetylsalicylic acid/aspirin. As evident, this drug – based on its activity under aerobic conditions within the system HA-ascorbate-Cu²⁺-NaOCl – can be classified as a potent scavenger of •OH radicals [48].

15.6   CONCLUSIONS

With the current understanding that free radicals can act as cell signaling or "messenger" agents it is likely that they also play a role in normal cellular function as well as various disease etiologies. Researchers are now making rapid progress in understanding the role of oxidative stress and nitrosative stress in cardiovascular diseases such as atherosclerosis, ischemia/reperfusion injury, restenosis and hypertension; cancer; inflammatory diseases such as acute respiratory distress syndrome (ARDS), asthma, inflammatory bowel disease (IBD), dermal and ocular inflammation and arthritis; metabolic diseases such as diabetes; and diseases of the central nervous system (CNS) such as amyotrophic lateral sclerosis (ALS), Alzheimer's, Parkinson's, and stroke. The increased awareness of oxidative stress related to disease and the need to measure the delicate balance that exists between free radicals and the given systems in regulating them has given rise to a demand for new research tools.

KEYWORDS

- **Osteoarthritis (OA)**
- **Polymorphonuclear leukocyte (PMNL)**
- **Rheumatoid arthritis (RA)**

ACKNOWLEDGMENT

The work was supported by the VEGA grant No. 2/0011/11 and the APVV grant No. 0351-1.

REFERENCES

1. Yan, D., Gao, C., and Frey, H. *"Hyperbranched Polymers Synthesis, Properties, and Applications*, John Wiley & Sons, New Jersey (2011).
2. Frechet, J. M. J. and Tomalia, D. A. *Dendrimers and other dendritic polymers*, John Wiley & Sons, UK (2011).
3. Hasanzadeh, M., Moieni, T., and Hadavi Moghadam, B. *Adv. Polym. Tech.*, **32**, 792 (2013).
4. Mishra, M. K. and Kobayashi, S. *Star and Hyperbranched Polymers*, Marcel Dekker, New York (1999).
5. Gates, T. S. and Odegard, G. M., Frankland, S. J. V., Clancy, T. C. *Compos. Sci. Technol.*, **65**, 2416 (2005).
6. Jikei, M. and Kakimoto, M. *Prog. Polym. Sci.*, **26**, 1233 (2001).
7. Gao, C. and Yan, D. *Prog. Polym. Sci.*, **29**, 183 (2004).
8. Voit, B. I. and Lederer, A. *Chem. Rev.*, **109**, 5924 (2009).
9. Kumar, A. and Meijer, E. W. *Chem. Commun.*, **16**, 1629 (1998).
10. Grabchev, I., Petkov, C., Bojinov, V., *Dyes. Pigments.*, **62**, 229 (2004).
11. Qing-Hua, C., Rong-Guo, C., Li-Ren, X., Qing-Rong, Q., and Wen-Gong, Z. *Chinese J. Struct. Chem.*, **27**, 877 (2008).
12. Kou, Y., Wan, A., Tong, S., Wang, L., and Tang, J. *React. Funct. Polym.*, **67**, 955 (2007).
13. Schmaljohann, D., Pötschke, P., Hässler, R., Voit, B. I., Froehling, P. E., Mostert, B., and Loontjens, J. A. *Macromolecules*, **32**, 6333 (1999).
14. Kim, Y. H. *J. Polym. Sci. A*, **36**, 1685 (1998).
15. Liu, G. and Zhao, M. *Iranian Polym. J.*, **18**, 329 (2009).
16. Inoue, K. *Prog. Polym. Sci.*, **25**, 453 (2000).
17. Seiler, M. *Fluid Phase Equilibr.*, **241**, 155 (2006).
18. Yates, C. R. and Hayes, W. *Eur. Polym. J.*, **40**, 1257 (2004).
19. Voit, B. *J. Polym. Sci. A*, **38**, 2505 (2000).
20. Nasar, A. S., Jikei, M., and Kakimoto, M. *Eur. Polym. J.*, **39**, 1201 (2003).
21. Froehling, P. E. *Dyes. Pigments.*, **48**, 187 (2001).
22. Jikei, M., Fujii, K., and Kakimoto, M. *Macromol. Symp.*, **199**, 223 (2003).
23. Radke, W., Litvinenko, G., and Müller, A. H. E. *Macromolecules*, **31**, 239 (1998).
24. Maier, G., Zech, C., Voit, B., and Komber, H. *Macromol. Chem. Phys.*, **199**, 2655 (1998).
25. Voit, B., Beyerlein, D., Eichhorn, K., Grundke, K., Schmaljohann, D., and Loontjens, T. *Chem. Eng. Technol.*, **25**, 704 (2002).
26. Frey, H. and Hölter, D. *Acta. Polym.*, **50**, 67 (1999).
27. Hölter, D., Burgath, A., and Frey, H. *Acta. Polymerica.*, **48**, 30 (1997).
28. Le, T.C., Todd, B. D., Daivis, P. J., and Uhlherr, A. *J. Chem. Phys.*, **130**, 074901 (2009).
29. Haj-Abed, M. Master Degree Thesis, McMaster University (2008).
30. Stiriba, S. E., Kautz, H., and Frey, H. *J. Am. Chem.*, **124**, 9698 (2002).
31. Zou, J., Zhao, Y., Shi, W., Shen, X., and Nie, K. *Polym. Advan. Technol.*, **16**, 55 (2005).
32. Zhu, L., Shi, Y., Tu, C., Wang, R., Pang, Y., Qiu, F., Zhu, X., Yan, D., He, L., Jin, C., and Zhu, B. *Langmuir*, **26**, 8875 (2010).
33. Zhou, Y., Huang, W., Liu, J., Zhu, X., and Yan, D. *Adv. Mater.*, **22**, 4567 (2010).

34. Dodiuk-Kenig, H., Lizenboim, K., Eppelbaum, I., Zalsman, B., and Kenig, S. *J. Adhes. Sci. Technol.*, **18**, 1723 (2004).

35. Burkinshaw, S. M., Froehling, P. E., and Mignanelli, M. *Dyes. Pigments.*, **53**, 229 (2002).

36. Mulkern, T.J. and Tan, N. C. *Polymer*, **41**, 3193 (2000).

37. Boogh, L., Pettersson, B., and Månson, J. A. E. *Polymer*, **40**, 2249 (1999).

38. Zhang, F., Chen, Y., Lin, H., and Lu, Y. *Color. Technol.*, **123**, 351 (2007).

39. Zhang, F., Chen, Y., Lin, H., Wang, H., and Zhao, B. *Carbohyd. Polym.*, **74**, 250 (2008).

40. Zhang, F., Chen, Y.Y., Lin, H., and Zhang, D. S. *Fiber. Polym.*, **9**, 515 (2008).

41. Zhang, F., Chen, Y., Ling, H., and Zhang, D. *Fiber. Polym.*, **10**, 141 (2009).

42. Hasanzadeh, M., Moieni, T., and Hadavi Moghadam, B. *J. Polym. Eng.*, **20**, 2191 (2013).

43. Khatibzadeh, M., Mohseni, M., and Moradian, S. *Color. Technol.*, **126**, 269 (2010).

44. Mahmoodi, R., Dodel, T., Moieni, T., and Hasanzadeh, M. *Polym. Res. J.*, **7**, 20, (2013).

45. Hua, Y., Zhang, F., Lin, H., and Chen, Y. *Textile Auxiliaries*, **20** (2008).

46. Zhang, F., Zhang, D., Chen, Y., and Lin, H. *Cellulose*, **16**, 281 (2009).

47. Burkinshaw, S. M., Mignanelli, M., Froehling, P. E., and Bide, M. J. *Dyes. Pigments.*, **47**, 259 (2000).

48. Gao, C., Xu, Y., Yan, D., and Chen, W. *Biomacromolecules*, **4**, 704 (2003).

49. Wu, D., Liu, Y., Jiang, X., He, C., Goh, S. H., and Leong, K. W. *Biomacromolecules*, **7**, 1879 (2006).

50. Kima, T. H., Cooka, S. E., Arote, R. B., Cho, M. H., Nah, J. W., Choi, Y. J., and Cho, C. S. *Macromol. Biosci.*, **7**, 611 (2007).

51. Reul, R., Nguyen, J., and Kissel, T. *Biomaterials*, **30**, 5815 (2009).

52. Goswami, A. and Singh, A. K. *React. Funct. Polym.*, **61**, 255 (2004).

53. Fang, J., Kita, H., and Okamoto, K. *Macromolecules*, **33**, 4639 (2000).

54. Seiler, M., Köhler, D., and Arlt, W. *Sep. Purif. Technol.*, **30**, 179 (2003).

55. Itoh, T., Hirai, K., Tamura, M., Uno, T., Kubo, M. and Aihara, Y. *J. Solid. State. Electr.*, **14**, 2179 (2010).

56. Tang, W., Huang, Y., Meng, W., Qing, F. L. *Eur. Polym. J.*, **46**, 506 (2010).

57. Hong, X., Chen, Q., Zhang, Y., and Liu, G. *J. Appl. Polym. Sci.*, **77**, 1353 (2000).

58. Shokrieh, M. M. and Rafiee, R. *Mater. Design.*, **31**, 790 (2010).

59. Tserpes, K. I. and Papanikos, P. *Composites B*, **36**, 468 (2005).

60. Arani, A. G., Rahmani, R., and Arefmanesh, A. *Physica E*, **40**, 2390 (2008).

61. Ruoff, R. S., Qian, D., Liu, and W. K. *C. R. Physique*, **4**, 993 (2003).

62. Guo, X., Leung, A. Y. T., He, X. Q., Jiang, H., and Huang, Y. *Composites B*, **39**, 202 (2008).

63. Xiao, J. R., Gama, B. A., and Gillespie Jr, J. W. *Int. J. Solid. Struct.*, **42**, 3075 (2005).

64. Ansari, R. and Motevalli, B. *Commun. Nonlinear. Sci. Numer. Simulat.*, **14**, 4246 (2009).

65. Li, C. and Chou, T. W. *Mech. Mater.*, **36**, 1047 (2004).

66. Natsuki, T. and Endo, M. *Carbon*, **42**, 2147 (2004).

67. Ansari, R., Sadeghi, F., and Motevalli, B. *Commun. Nonlinear. Sci. Numer. Simulat.*, **18**, 769 (2013).

68. Alisafaei, F. and Ansari, R. *Comput. Mater. Sci.*, **50**, 1406 (2011).

69. Natsuki, T., Ni, Q.Q., and Endo M. *Appl. Phys. A*, **90**, 441 (2008).

70. Joshi, U. A., Sharma, S. C., and Harsha, S. P. *Proc. IMechE, Part N: J. Nanoeng. Nanosys.*, **225**, 23 (2011).

71. Shokrieh, M. M. and Rafiee, R. *Mech. Compos. Mater.*, **46**, 155 (2010).

72. Sadus, R. J. "*Molecular Simulation of Fluids: Algorithms and Object- Orientation*", Elsevier, Amsterdam, 1999.

73. Allen, M. P. and Tildesley, D. J. *"Computer simulation in chemical physics"*, Dordrecht Kluwer Academic Publishers, Dordrecht (1993).
74. Haile, J. M. *"Molecular dynamics simulation: elementary methods"*, Wiley, New York (1997).
75. Liu, W. K., Karpov, E. G., and Park, H. S. *"Nano Mechanics and Materials: Theory, Multiscale Methods and Applications"*, John Wiley & Sons, New York (2006).
76. Rapaport, D. C. *"The Art of Molecular Dynamics Simulation"*, Cambridge University Press, New York (1995).
77. Aerts, J. *Comput. Theor. Polym. Sci.*, **8**, 49 (1998).
78. Lescanec, R. L. and Muthukumar, M. *Macromolecules*, **23**, 2280 (1990).
79. Le, T.C., Todd, B. D., Daivis, P. J., and Uhlherr, A. *J. Chem. Phys.*, **130**, 074901 (2009).
80. Lyulin, A. V., Adolf, D. B., and Davies, G. R. *Macromolecules*, **34**, 3783 (2001).
81. Konkolewicz, D., Thorn-Seshold, O., and Gray-Weale, A. *J. Chem. Phys.*, **129**, 054901 (2008).
82. de Gennes, P. G. and Hervet, H. J. *J. Phys. Lett. (Paris)*, **44**, L351 (1983).
83. Maiti, P. K., Cagın, T., Wang, G. F., and Goddard, W. A. *Macromolecules*, **37**, 6236 (2004).
84. Cagin, T., Wang, G., Martin, R., Breen, N., and Goddard, W. A. *Nanotechnology*, **11**, 77 (2000).
85. Cagin, T., Wang, G., Martin, R., Zamanakos, G., Vaidehi, N., Mainz, D.T., and Goddard, W. A. *Comput. Theor. Polym. Sci.*, **11**, 345 (2001).
86. Lue, L. *Macromolecules*, **33**, 2266 (2000).
87. Le, T. C. Ph.D. Degree Thesis, Swinburne University of Technology (2010).
88. Zacharopoulos, N. and Economou, I. G. *Macromolecules*, **35**, 1814 (2002).
89. Lee, I., Athey, B. D., Wetzel, A. W., Meixner, W., and Baker, J. R. *Macromolecules*, **35**, 4510 (2002).
90. Hana, M., Chen, P., and Yang, X. *Polymer*, **46**, 3481 (2005).
91. Laferla, R. *J. Chem. Phys.*, **106**, 688 (1997).
92. Avila-Salas, F., Sandoval, C., Caballero, J., Guiñez-Molinos, S., Santos, L. S., Cachau, R. E., and González-Nilo, F. D. *J. Phys. Chem. B*, **116**, 2031 (2012).
93. He, X., Liang, H., and Pan, C. *Macromol. Theory Simul.*, **10**, 196 (2001).
94. Neelov, I. M. and Adolf, D. B. *J. Phys. Chem. B*, **108**, 7627 (2004).
95. Bosko, J. T. and Prakash, J. R. *J. Chem. Phys.*, **128**, 034902 (2008).
96. Dalakoglou, G. K., Karatasos, K., Lyulin, S. V., and Lyulin, A. V. *J. Chem. Phys.*, **129**, 034901 (2008).
97. Konkolewicz, D., Gilbert, G. R., and Gray-Weale, A. *Phys. Rev. Lett.*, **98**, 238301 (2007).
98. Sheridan, P. F., Adolf, D. B., Lyulin, A. V., Neelov, I., and Davies, G. R. *J. Chem. Phys.*, **117**, 7802 (2002).

CHAPTER 16

POLYVINYLCHLORIDE ANTIBACTERIAL PRE-TREATED BY BARRIER PLASMA

IGOR NOVÁK, JÁN MATYAŠOVSKÝ, PETER JURKOVIČ, MARIÁN LEHOCKÝ, ALENKA VESEL, and LADISLAV ŠOLTÉS

CONTENTS

16.1 INTRODUCTION

A multistep physicochemical approach making use of plasma technology combined with wet chemistry has fueled considerable interest in delivery of surface-active anti-adherence materials. In the first step of the approach, concerning an inherent lack of befitting functional groups on pristine substrate, plasma treatment at low temperature and atmospheric pressure has been substantiated to be productive in yielding reactive entities on the surface [1,5]. The highlights the functionality of the adopted multi-step physicochemical approach to bind polysaccharide species onto the medical-grade PVC surface. DCSBD plasma is capable of raising roughness, surface free energy, and introducing oxygen-containing functionalities anchored onto the surface. A structured poly(acrylic acid) brush of high graft density is synthesized using surface-initiated approach to further improve hydrophilicity and develop a stable brush-like assembly to yield a platform for biomolecular binding. In vitro bacterial adhesion and biofilm formation assays indicate incapability of single chitosan layer in hindering the adhesion of Staphylococcus aureus bacterial strain. Chitosan could retard Escherichia coli adhesion and plasma treated and graft copolymerized samples are found effective to diminish the adherence degree of Escherichia Coli.

A new modification method using plasma technology combined with wet chemistry represents an efficient way in delivery of surface-active anti-adherence materials [1-4]. The atmospheric pressure electric discharge plasma has been substantiated to be productive in yielding reactive entities on the surface [5,6]. However, the need for treatment duration to a few seconds remains a pressing obstacle to extensive applications of this type of plasma [7]. A novel technology coined as diffuse coplanar surface barrier discharge (DCSBD) has been developed [8], which enables the generation of a uniform plasma layer under atmospheric pressure with a high surface power density in the very close contact of modified polymer.

16.2 EXPERIMENTAL

- *Materials*: PVC pellets, extrusion medical-grade RB1/T3M of 1.25 g·cm⁻³ density, were obtained from ModenPlast (Italy) and used as received. Pectin from apple, (BioChemika, with esterification of 70-75%), acrylic acid (AA) (99.0%, anhydrous), and *N*-(3-dimethyl aminopropyl)-*N'*-ethyl carbodiimide hydrochloride (EDAC, 98.0%) were supplied by Fluka (USA). Chitosan from crab shells with medium molecular weight and deacetylation degree of 75-85%.
- *Plasma Modification*: It was implemented in static conditions by DCSBD plasma technology (Figure 1) of laboratory scale with air as the gaseous medium at atmospheric pressure and room temperature. A schematic profile of the plasma system is given in Scheme 1. It basically comprises a series of parallel metallic electrodes inset inside a ceramic dielectric located in a glass chamber which allows the carrier gases to flow. All samples were treated on both sides with plasma power of 200 W for 15 sec.

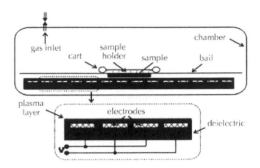

SCHEME 1 Scheme of DSBD plasma source.

For grafting by AA PVC substrates were immersed into spacer solutions containing 10 vol.% AA aq. solution. The reaction was allowed to proceed for 24 hr at 30°C. *PAA grafted PVC samples were immersed into EDAC aq. solution at 4°C for 6 hr* in order to activate the carboxyl groups on the surface. The highly active key intermediate, O-acylisourea, is produced having potential to react with reducing agents. Subsequently, they were transferred to chitosan and kept there for 24 hr at 30°C.

Sample 1 – pristine PVC, sample 2 –PVC treated by DCSBD plasma, sample 3 – PVC treated by plasma and grafted by AA, sample 4 – PVC treated by plasma, AA and chitosan, sample 5 – PVC treated by plasma AA, chitosan and pectin.

Scanning electron microscopy (SEM) was carried out on VEGA II LMU (TESCAN) operating in the high vacuum/secondary electron imaging mode at an accelerating voltage of 5–20 kV. Bacterial adhesion and biofilm experiments were performed using gram-positive (*S. Aureus* 3953) and gram-negative (*E. Coli* 3954) bacteria. The circular shape specimens (d ≈ 8mm) were cut from the pristine and modified PVC samples before further investigation. After 24 hr incubation at 37°C *under continuous shaking at 100 rpm. The bacteria adhered on the surface of the specimens were removed by vigorous shaking of the test tube at 2000 rpm for 30 sec and quantified by serial dilutions and spread plate technique.*

16.3 DISCUSSION AND RESULTS

16.3.1 SURFACE ENERGY

Table 1 includes the contact angle values of deionized water (θ_w) recorded on different samples. Each sample has been designated by a number from 1 to 5 whose notation is inserted in the title of Table 1. Based on the given data, sample 1 exhibits a hydrophobic characteristic which after being treated by plasma, an evident change in θ_w arises and hydrophilicity ascends as anticipated. This trend continues as to sample 3 on which polyacrylic acid (PAA) chains are grafted where more hydrophilic propensity is shown inferred from θ_w value. The elevated hydrophilicity upon multistep modifications is assumed to come from the inclusion of superficial hydrophilic entities. The hydrophilicity then decreases as polysaccharides are coated onto the surface, though is well higher than that of sample 1, as the inherent hydrophilicity of chitosan is beyond doubt. Furthermore, sample 5 exhibits higher wettability than sample 4 implying a more effective binding of chitosan onto the surface, as remarked in other efforts as well. The hydrophilicity then decreases as polysaccharides are coated onto the surface, though is well higher than that of sample 1, as the inherent hydrophilicity of chitosan is beyond doubt. Furthermore, sample 5 exhibits higher wettability than sample 4 implying a more effective binding of chitosan onto the surface, as remarked in other efforts as well.

TABLE 1 Contact angle analysis results of different specimens using deionized water (W), ethylene glycol (E), diiodomethane (D), and formide (F) as wetting agent. Sample 1: pristine control; Sample 2: plasma treated; Smple 3: PAA grafted; Sample 4: chitosan coated; Sample 5: chitosan/pectin coated (mean ± standard deviation)

Specimen	θ_n(°)	θ_E(°)	θ_D(°)	θ_F(°)	$\gamma^{LW/AB}$ (mj/m²)	$\gamma^{+LW/AB}$ (mj/m²)	$\gamma^{AB}_{LW/AB}$ (mj/m²)	$\gamma^{LW}_{LW/AB}$ (mj/m²)	$\gamma^{tot}_{LW/AB}$ (mj/m²)	$\gamma^{a)}{}_{Wa}$ (mj/m²)	$\gamma^{b)}{}_{KN}$ (mj/m²)	$\gamma^{c)}{}_{LN}$ (mj/m²)
Sample 1	85.9 (±2.5)	60.5 (±3.0)	43.5 (±3.5)	64.2 (±6.0)	5.1	0.0	1.0	37.8	38.8	37.8	33.3	33.6
Sample 2	64.9 (±3.0)	49.4 (±4.0)	36.2 (±5.5)	51.0 (±6.0)	24.9	0.5	6.7	41.5	48.2	41.5	40.4	40.7
Sample 3	46.5 (±4.0)	51.3 (±5.5)	38.0 (±5.0)	7.7 (±4.5)	62.9	2.7	26.1	40.6	66.7	51.9	43.1	43.4
Sample 4	63.7 (±5.5)	43.4 (±3.0)	28.2 (±2.5)	44.9 (±5.0)	22.2	0.3	4.9	45.0	49.9	45.0	42.8	43.0
Sample 5	50.5 (±3.5)	40.0 (±2.5)	31.5 (±4.5)	31.0 (±3.5)	42.2	0.6	10.5	43.6	54.1	50.0	46.4	46.6

a) Surface free energy value according to Wu equation of state [33]. b) Surface free energy value according to Kwok–Neumann model [33].
c) Surface free energy value according to Li-Neumann Model.

To further explore the physicochemical parameters of the examined surfaces, an extensively used theory, Lifshitz-van der Waals/acid-base (LW/AB), has been exploited for free surface energy evaluation whose outputs with reference to diiodomethane, ethylene glycol, and deionized water as wetting liquids are supplied in Table 1. Sample 1 exhibits a basic character ($\gamma^- > \gamma^+$) as proposed by the data, even though acidity or basicity of neat PVC is yet controversial.

This increase is principally assisted by the polar (acid-base) component (γ^{AB}), rather than the apolar one (γ^{LW}), implying an incorporation of superficial polar oxygen-containing entities thanks to the air plasma treatment. A significant rise in γ^{tot} and γ^{AB} values is noticed for sample 3, in comparison with samples 1 and 2, indicative of the presence of carboxyl-containing units on the surface. As for samples 4 and 5, a reduction in γ^{AB} and γ^{tot} values is observed compared to sample 3, however, their γ^{tot} values rise above that of sample 1. The minimum values of θ_E and θ_F are found for sample 5 which reflect that the surface is seemingly coated by alcoholic and amine containing moieties which in fact points to the more efficient binding of chitosan when compared to sample 4.

16.3.2 SURFACE MORPHOLOGY

The surface topography of samples 1–5 investigated by SEM as a common surface qualitative technique are presented in Figure 2. Sample 1 shows a level and uniform morphology which goes through a significant alteration ensuing the plasma treatment taking on an etched pattern with an unevenly shaped texture. The generated morphology is favorable for next coupling processes due to an enhanced surface area and roughness. The developed pattern on sample 2 is indeed, an outcome of the competing functionalization and ablation phenomena which brings on a reorganization of the surface microstructure.

Sample 1   Sample 2

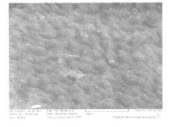

Sample 3

FIGURE 1 *Continued*

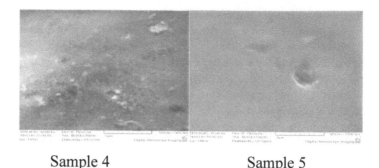

Sample 4   Sample 5

FIGURE 1   SEM micrographs of samples 1–5 taken at 3×10^4 magnification.

The incident of the ablation is validated by gravimetric analysis where a weight loss of 4 $\mu g \cdot cm^{-2}$ has been observed due to the plasma treatment for 15 sec implying an approximate etching rate of 2 nm/s in terms of the used PVC grade density. Based on the sample 3 micrograph, PAA chains develop superficial domains of submicron dimension and brush-like features are then recognizable on the surface. As the grafting moves forward, clustering takes place because of the domains size growth. An additional compelling factor in controlling the surface microstructure is the grafting mechanism which is actually initiated by generated surface radicals.

16.3.3   SURFACE CHEMISTRY XPS ANALYSIS

XPS, with a probe depth measuring around 5 nm, has been put to use to more thoroughly monitor the bearings of the surface modifications by picking up a quantitative perception into the surface elemental composition. The recorded survey spectra along with the corresponding surface atomic compositions and ratios of samples 1-5 are all provided in Figure 4. Carbon (C), oxygen (O), chlorine (Cl), and silicon (Si) elements are found on the sample 1 surface whose composition and elemental ratios are presented in the legend of the respective graph. The Cl2p atomic content is substantially lower than the amount found for a neat PVC containing no additives which refers to the existence of several additives and also X-ray degradation. The same rationale accounts for the considerable amount of O1s detected in sample 1 which is not a typical element in standard PVC.Upon binding chitosan on the surface (sample 4), pronounced changes appear in the surface chemistry, as O1s content and O/C fraction increase and also N1s signal emerges, while Cl2p and Si2p bands abate due to the surface coverage by polysaccharide species. This trend yet continues for sample 5 as higher O1s and N1s as well as O/C and N/C atomic rations are detectable compared to sample 4 giving support to the notion that chitosan can be more stably, i.e. in higher quantity, attached onto the surface when layered along with pectin. In other words, use of pectin can promote the quality of chitosan binding.

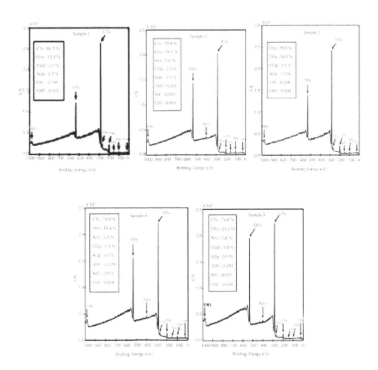

FIGURE 2   XPS survey-scan spectra of samples 1-5 along with atomic compositions.

BACTERIAL ADHESION AND BIOFILM ASSAY

The most crucial step of the biofilm formation is bacterial adhesion considered as a so-phisticated topic in biointerface science whose plenty of aspects have not yet been well conceived. As a matter of fact, adhesion phenomenon is an interplay of myriad factors. Figure 5 shows the histograms of bacterial adhesion extent for samples 1-5 after 24 h in-cubation. As Regards the adherence degree of *S. aureus* onto the samples 2-4, no reduc-tion is evident in the number of viable adhered colonies, compared to sample 1, signify-ing an inability of the modifications in hampering the *S. aureus* adhesion to the surface. From sample 1 to 3, both hydrophilicity and roughness rise, as remarked earlier, and then decrease in the case of samples 4 and 5. The adhesion degrees vary with a similar trend as well. Considering sample 5, it is inferred that chitosan/pectin assembly imparts biocidal effects against *S. aureus*. Chitosan single layer and chitosan/pectin multilayer restrain the adherence degree by 50% and 20%, respectively. Chitosan/pectin multilayer is found to be effective against both gram-positive and gram-negative strains which can be translated as a higher quality of chitosan coating when it is applied along with pectin.

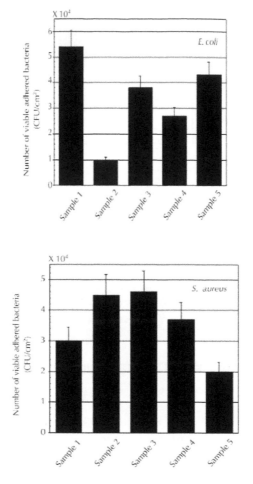

FIGURE 3   Histograms of bacterial adhesion degree for samples 1-5 after 24 h incubation against two microorganisms.

16.4   CONCLUSION

The DCSBD plasma is capable of raising roughness, surface free energy, and introducing oxygen-containing functionalities anchored onto the PVC surface. A structured PAA brush of high graft density is synthesized using surface-initiated approach to further improve hydrophilicity and develop a stable brush-like assembly to yield a platform for biomolecular binding. In vitro bacterial adhesion and biofilm formation assays indicate incapability of single chitosan layer in hindering the adhesion of *S. aureus* bacterial strain, while up to 30% reduction is achieved by chitosan/pectin layered assembly. On the other hand, chitosan and chitosan/pectin multilayer could retard *E. coli* adhesion by 50% and 20%, respectively. Furthermore, plasma treated and graft copolymerized samples are also found effective to diminish the adherence degree of *E. coli.*

ACKNOWLEDGMENT

This paper was processed in the frame of the APVV project No. APVV-351-10 as the result of author's research at significant help of APVV agency Slovakia.

REFERENCES

1. Valko, M., Rhodes, C. J., Moncol, J., Izakovic, M., and Mazur, M. *Chem.-Biol. Interact.*, **160**, 1–40 (2006).
2. Harman, D. *J. Gerontol.*, **11**, 298–300 (1956).
3. Valko, M., Leibfritz, D., Moncol, J., Cronin, M. T. D., Mazur, M., Telser, J., *Int. J. Biochem. Cell Biol.*, **39**, 44–84 (2007).
4. http://www.niams.nih.gov/Health_Info/Rheumatic_Disease/graphics/joint_vert.gif.
5. Servaty, R., Schiller, J., Binder, H., and Arnold, K. *Int. J. Biol. Macromol.*, **28**, 121–127 (2001).
6. Http://www.nature.com/ncprheum/journal/v2/n7/fig_tab/ncprheum0216_F1.html#figure-title .Chen, F. H., Rousche, K. T., and Tuan, R. S. *Nat. Clin. Prac. Rheumatol.*, **2**, 373–382 (2006).
7. Praest, B. M., Greiling, H., and Kock, R. *Carbohydr. Res.*, **303**, 153–157 (1997).
8. Oates, K. M. N., Krause, W. E., and Colby, R. H. *Mat. Res. Soc. Symp. Proc.*, **711**, 53–58 (2002).
9. Rovenská, E. *Rheumatologia*, **15**, 57–64 (2001).
10. Day, R. O., McLachlan, A. J,. Graham, G. G., and Williams, K. M. *Clin. Pharmacokin.*, **36**, 191–210 (1999).
11. Kogan, G., Šoltés, L., Stern, R., and Mendichi, R. "Hyaluronic acid: A biopolymer with versatile physico-chemical and biological properties" in the book "*Handbook of Polymer Research: Monomers, Oligomers, Polymers and Composites*, Chapter 31", R. A. Pethrick, A. Ballada, and G. E. Zaikov (Eds.), Nova Science Publishers, New York, pp. 393–439 (2007).
12. Ryabina, V. R., Vasyukov, S. E., Panov, V. P., and Starodubtsev, S. G. *Khim.-Farm. Zh.* **21**, 142–149 (1987).
13. Hardingham, T. "Solution properties of hyaluronan" in the book *"Chemistry and Biology of Hyaluronan"*, pp 1–19, H. G. Garg and C. A. Hales (Eds.) Elsevier Press, Amsterdam (2004).
14. Gibson, J. S., Milner, P. I., White, R., Fairfax, T. P., and Wilkins, R. *J. Pflugers Arch.* **455**, 563–573 (2008).
15. Fang, Y. Z., Yang, S., and Wu, G. *Nutrition*, **18**, 872–879 (2002).
16. Šoltés, L., and Kogan, G. "Impact of transition metals in the free-radical degradation of hyaluronan biopolymer" in the book *"Kinetics & Thermodynamics for Chemistry & Biochemistry*, Vol. 2", pp. 181–199, E. M. Pearce, G. E. Zaikov, and G. Kirshenbaum, (Eds.), Nova Science Publishers, New York 2009, *ibid Polym. Res. J.*, **2**, 269–288 (2008).
17. Milner, P. I., Wilkins, R. J., and Gibson, J. S. *Osteoarthr. Cartilage*, **15**, 735–742 (2007).
18. Hawkins, C. L. and Davies, M. J. *Free Radic. Biol. Med.*, **21**, 275–290 (1996).
19. Kennett, E. C. and Davies, M. J. *Free Radic. Biol. Med.*, **47**, 389–400 (2009).
20. Hawkins, C. L. and Davies, M. J. *Biochem. Soc. Trans.*, **23**, S248 (1995).
21. Kogan, G., Šoltés, L., Stern, R., Schiller, J., and Mendichi, R. "Hyaluronic Acid: Its Function and Degradation in In Vivo Systems", in the book *"Studies in Natural Products Che-*

mistry, Vol. 35 Bioactive Natural Products, Part D", pp. 789–882, Atta-ur-Rahman (Ed.), Elsevier, Amsterdam (2008).

22. Myint, P., Deeble, D. H., Beaumont, P. C., Blake, S. M., and Phyllips, G. O. *Biochim. Biophys. Acta*, **925**, 194–202 (1987).

23. Presti, D. and Scott, J. E. *Cell Biochem. Func.*, **12**, 281–288 (1994).

24. Grootveld, M., Henderson, E. B., Farrell, A., Blake, D. R., Parkes, H. G., and Haycock, P. *Biochem. J.*, **273**, 459–467 (1991).

25. Niedermeier, W. and Griggs, J. H., *J. Chronic Dis.*, **23**, 527–536 (1971).

26. "Transport and Storage of Metal Ions in Biology", in the book "*Biological Inorganic Chemistry: Structure and Reactivity*". I. Bertini, H. B. Gray, E. I. Stiefel, and J. S. Valentine (Eds.), University Science Books, Sausalito California, pp. 57–77 (2007).

27. Halliwell, B. and Gutteridge, J. M. C. *Methods Enzymol.*, **186**, 1–85 (1990).

28. Wong, S.F., Halliwell, B., Richmond, R., and Skowroneck, W. R. *J. Inorg. Biochem.*, **14**, 127–134 (1981).

29. Hitchon, C. A. and El-Gabalawy, H. S. *Arthritis Res. Ther.*, **6**, 265–278 (2004).

30. http://www.oxisresearch.com/library/oxidative_stress.shtml.

31. http://books.google.sk/books?id=a52CE0vWRXQC&pg=PA321&lpg=PA321&d q=standard+reduction+potential+H2O2/%E2%80%A2OH&source=bl&ots=YJL 8hdl4Wh&sig=ChpmRnhvt--GEuyfkbotmWkvCVg&hl=sk&ei=aTxcSvu9HoyF_ AbQl6HwDA&sa=X&oi=book_result&ct=result&resnum=7.

32. http://www.gaiaresearch.co.za/hydroperoxide.html.

33. Martínez-Cayuela, M. *Biochimie*, **77**, 147–161 (1995).

34. **Dröge, W.** *Physiol. Rev.*, **82**, 47–95 (2002).

35. Halliwell, B., Hoult, J. R., and Blake, D. R. *FASEB J.*, **2**, 2867–2873 (1988).

36. Schiller, J., Fuchs, B., Arnhold, J., and Arnold, K. *Curr. Med. Chem.*, **10**, 2123–2145 (2003).

37. Brandt, K. D. "*An Atlas of Osteoarthritis*" The Parthenon Publishing Group Inc., New York, USA (2001).

38. http://en.wikipedia.org/wiki/inflammation.

39. Shukla, N., Maher, J., Masters, J., Angelini, G. D., and Jeremy, J. Y. *Atherosclerosis*, **187**, 238–250 (2006).

40. Weissberger, A., LuValle, J. E., and Thomas, D. S. Jr. *J. Am. Chem. Soc.*, **65**, 1934–1939 (1943).

41. Khan, M. M. and Martell, A. E. *J. Am. Chem. Soc.*, **89**, 4176–4185 (1967).

42. Fisher, A. E. O. and Naughton, D. P. *Med. Hypotheses*, **61**, 657–660 (2003).

43. Fisher, A. E. O. and Naughton, D. P. *Nutr. J.*, **3**, 1–5 (2004).

44. Fisher, A. E. O. and Naughton, D. P. *Curr. Drug Deliv.*, **2**, 261–268 (2005).

45. Šoltés, L., Stankovská, M., Brezová, V., Schiller, J., Arnhold, J., Kogan, G., and Gemeiner, P. *Carbohydr. Res.*, **341**, 2826–2834 (2006).

46. Valachová, K., Kogan, G., Gemeiner, P., and Šoltés, L. "Hyaluronan degradation by ascorbate: Protective effects of manganese(II) chloride" in the book "Kinetics & Thermodynamics for Chemistry & Biochemistry, Vol. 2", pp. 201–215, E. M. Pearce, G. E. Zaikov, and G. Kirshenbaum (Eds.), Nova Science Publishers, New York 2009, *ibid Polym. Res. J.*, **2**, 269–288 (2008).

47. Jiang, D., Liang, J., Noble, P. W., *Annu. Rev. Cell Dev. Biol.*, **23**, 435–461 (2007).

48. Stankovská, M., Arnhold, J., Rychlý, J., Spalteholz, H., Gemeiner, P., Šoltés, L. *Polym. Degrad. Stabil.*, **92**, 644–652 (2007).

INDEX